Gebreselassie Redae
Dereje Asefa
Solomon Habtu

Enhancing production and quality of groundnut in Northern Ethiopia

Gebreselassie Redae
Dereje Asefa
Solomon Habtu

Enhancing production and quality of groundnut in Northern Ethiopia

LAP LAMBERT Academic Publishing

Impressum / Imprint
Bibliografische Information der Deutschen Nationalbibliothek: Die Deutsche Nationalbibliothek verzeichnet diese Publikation in der Deutschen Nationalbibliografie; detaillierte bibliografische Daten sind im Internet über http://dnb.d-nb.de abrufbar.
Alle in diesem Buch genannten Marken und Produktnamen unterliegen warenzeichen-, marken- oder patentrechtlichem Schutz bzw. sind Warenzeichen oder eingetragene Warenzeichen der jeweiligen Inhaber. Die Wiedergabe von Marken, Produktnamen, Gebrauchsnamen, Handelsnamen, Warenbezeichnungen u.s.w. in diesem Werk berechtigt auch ohne besondere Kennzeichnung nicht zu der Annahme, dass solche Namen im Sinne der Warenzeichen- und Markenschutzgesetzgebung als frei zu betrachten wären und daher von jedermann benutzt werden dürften.

Bibliographic information published by the Deutsche Nationalbibliothek: The Deutsche Nationalbibliothek lists this publication in the Deutsche Nationalbibliografie; detailed bibliographic data are available in the Internet at http://dnb.d-nb.de.
Any brand names and product names mentioned in this book are subject to trademark, brand or patent protection and are trademarks or registered trademarks of their respective holders. The use of brand names, product names, common names, trade names, product descriptions etc. even without a particular marking in this works is in no way to be construed to mean that such names may be regarded as unrestricted in respect of trademark and brand protection legislation and could thus be used by anyone.

Coverbild / Cover image: www.ingimage.com

Verlag / Publisher:
LAP LAMBERT Academic Publishing
ist ein Imprint der / is a trademark of
OmniScriptum GmbH & Co. KG
Heinrich-Böcking-Str. 6-8, 66121 Saarbrücken, Deutschland / Germany
Email: info@lap-publishing.com

Herstellung: siehe letzte Seite /
Printed at: see last page
ISBN: 978-3-659-37975-8

Table of Contents

List of tables

List of figures

Acknowledgment

My special thanks and sincere gratitude goes to my thesis advisor Dr. Dereje Assefa for his diligent guidance, encouragement, supervision and constructive criticism from proposal inception to final write up of this manuscript and allowing research fund from the NORAD project III in which he is coordinating. My special thanks goes also to Dr.Solomon Habtu for his valuable support in supervision and constructive criticism from proposal development to final write up of this manuscript. The financial support from Irish Aid (Operational Research) project, which was used to cover the expense for titution fee and research fund, is greatly acknowledged.

I am great full to Tigray Agricultural Research Institute (TARI) and Mekelle Agricultural Research Center (MARK) for allowing me to follow the postgraduate program and for all materials and administrative support provided during the course of the study.

I would also like to thank Abergelle Agricultural Research Center staff members for their encouragement and support while executing my thesis work. Special thanks goes to Birhanu Sibhatu , Daniel Desta and G/medhen G/meskel for their dedicated support, moral and hospitality during my stay at Abergelle.

Deepest thanks are extended to Niguse Tesfa mickael, Tesfay Teklehaimanot, Bereket Haileselassie and Mulubrhan kifle for devoting their time in supporting me technically, for their constructive comments while I was writing this manuscript.

Special thanks goes to, Mehari Asfaw, Ligealem, Girmay hailu, Birhanu Abreha, Addis Abreha, Amina Abderkadire and Hagoes W/gebrealsI as well as all Crop case team of MARC for their encouragement , technical advice and material help for the success of my study.

.

Abrivations

ANOVA- Analysis of Varianc

CSA- Central Statistics Authority

DAP – Diammonium Phosphate

DMRT – Duncan's Multiple Range Test

EIAR – Ethiopian Agricultural Research Institute

EC- electrical conductivity

EU – European Union

FAO - Food and Agricultural Organization

FDA – Food and Drug Administration

FMP- Fused Magnesium Phosphate

FYM- Farm Yard Manure

GAP- Good Agricultural Practice

HACCP - Hazard Analysis Critical Control Point

ICRISAT- International Crop Research Institute for Semi Arid Tropics

ppb – parts per billion

SSP—single super phosphate

UNIDO -United Nations Industrial Development Organization

Abstract

In dryland agriculture degradation of natural resources, declining soil fertility ,moisture stress, and diseases and pests, are the major groundnut production constraints. Besides Aspergillus infection and subsequent contamination of groundnut with aflatoxin is a major limitation in the study area (Tanqua Abergel,Tigray). This study was executed to evaluate the effect of DAP fertilizer and gypsum application, tied ridging and supplementary irrigation on groundnut productivity and level of Aspergillus infection. The experiment was laid out in Randomized Complete Block Design with three replications in two sites. DAP as a source of P, and gypsum as source of Ca, were applied at planting and pod setting stages, respectively. While tied ridging and supplementary irrigation were applied at early flowering and during cessation of rainfall, respectively. Phenological, yield, and yield components data and severity of Aspergillus on sampled kernels were recorded. The analysis of variance indicated that the integrated agronomic management options showed a significant positive effect on plant height, pods/per plant, dry biomass weight, dry pod weight, kernel seed yield, 100 seed weight and reduction of Aspergillus flavus infection on groundnut at both experimental sites. The highest kernel yield (980kg/ha) and minimum Aspergillus flavus infection (3%) were recorded in management options where supplementary irrigation + Tied ridge + Fertilizer and Supplementary irrigation + tied ridging were practiced at Hadinet, respectively. The lowest yield (290kg/ha) and highest Aspergillus flavus (17.3%) was recorded in the control. In Lemlem experimental site the highest kernel yield (959kg/ha) and minimum Aspergillus flavus infection (4.3%) was recorded in gypsum + supplementary irrigation combination. While lowest yield (485kg/ha) and high (19.3%) Aspergillus infection was recorded in control.

<u>Key words</u>: Agronomic practices; Aspergillus infection; groundnut:

Chapter one: Introduction

1.1Background

Groundnut (*Arachis hypogaea* L.) belongs to subtribe *Stylosanthinae* of tribe *Aeschynomeneae* of family *Leguminosae*. Groundnut is an allotetraploid (2n=4x=40) species which likely evolved from two diploids in section Arachis (Kochert *et al.*, 1996).Groundnut is the sixth most important oilseed crop in the world. Groundnut kernels are consumed directly as a raw, roasted or boiled or oil extracted from the kernel is used as culinary oil. The crop is also used as animal feed and industrial raw material. The multiple use of groundnut plant makes it an excellent cash crop for domestic markets as well as for foreign trade in several developing and developed countries. The oil content of the crop has been well appreciated and documented. It is the world's 4th most important source of edible oil (Bhatia et al., 2006). Groundnut seed contains high quality edible oil (50%), easily digestible protein (25%) and 20% carbohydrates (Bhatia et al., 2006).

Groundnut is grown on 26.4 million ha worldwide with a total production of 37.1 million metric ton and an average productivity of 1.4 metric ton ha^{-1} (FAO, 2003). Developing countries constitute 97% of the global area and 94% of the global groundnut production (Nigam et al., 2004). The production of groundnut is concentrated in Asia and Africa, with 56% and 40% of the global area and 68% and 25% of the global production, respectively (Nigam et al., 2004).

Groundnut is relatively recent to Ethiopia. It was introduced to northern Ethiopia by the Portuguese in the 17th century, and later through the Arab influence to south eastern part of the country (Brereton, 1980). It grows in the warmer regions of eastern and northern part of the country. The total area occupied by groundnut in Ethiopia was estimated to be 41, 578.79ha, (CSA, 2010).

Groundnut is grown in well drained sandy loam or sandy clay loam soil with pH ranging from 6.5 – 7 with high fertility. The loss of pod is usually high in heavier soils. An optimum soil temperature for good germination of groundnut is 30 °c. Low temperature at sowing delays germination and increases seed and seedling disease. The productivity is restrained by drought stress, use of low levels of inputs by smallholder farmers in marginal dry land areas, high incidence of foliar fungal diseases, and attack by insect pests. Several diseases and insect pests causing large losses in both yield and quality of seeds have been reported in groundnut crop. According to CSA (2009/10), agricultural sample survey, the productivity of groundnut in Ethiopia is 1.197 t/ha which is very low compared with the major groundnut producing countries. Average yields range from 2.1 tons/ha to 3.1 tons/ha (Steven and Luz, 2008).

Aflatoxin contamination has a significant role in groundnut production. Aflatoxins toxic fungal metabolites mainly produced by *Aspergillus flavus* and *A. parasiticus*. Contamination of groundnut with aflatoxin occurs under pre harvest, post harvest handling and storage conditions. The main factors leading to aflatoxin contamination include poor cultural practices; use of damaged and loose shelled kernels as seed and delayed harvesting after physiological maturity aggravates biological and physical effects of aflatoxins (UNIDO, 2010).

Aspergilli are group of fungi exhibiting immense ecological and metabolic diversity. These include notorious pathogen such as *Aspergillus fulvus*, which produces aflatoxin, one of the most potent, naturally occurring carcinogenic compounds known to man. It has been reported that aflatoxin in food and feed crops are almost entirely produced by Aspergillus spp (Bennett, 2010).

Besides, research findings clearly stated that from the known Aspergillus spp, *Aspergillus flavus* is the most widely reported food born fungus outside north – temperate areas (Bennett, 2010).

The development and contamination of the fungus in poorly managed fields is more serious. Timely planting, adequate fertility, good weeding and insect control, supplementary irrigation, suitable plant population and hybrid selection considerably reduce aflatoxin contamination (Erick, 2009).

The association of mycotoxins with consumption of contaminated groundnut and other food lead many importing countries, including the EU countries, to enact regulations establishing maximum levels for aflatoxins in groundnut and groundnut products (Steven and Rios, 2008).

According to FDA 25% of all the foods produced in the world are contaminated with aflatoxins. To reduce public health risk from consumption of contaminated food and feed, constant monitoring and of the toxins in food and feed become essential along with commodity value chain improvement (UNIDO, 2010).

The infection of *A. flavus* can be minimized to a level where it can not cause significant damage through the application of different pre- harvest management practices. According to Doner (2007), the risk to infection and contamination of aflatoxin can be reduced substantially with proper irrigation supply particularly from flowering to pod setting stages, soil amendments; application of proper and rapid drying after harvest.

Despite its importance as a food and cash crop in Ethiopia in general and study area in particular the yield and quality of groundnut is low which is mainly affected by stress problem especially at late growth stage of the crop. In addition, little work has been done on the extent of integrated management options which enhances production and productivities of groundnut and improving on its quality. Moreover, late onset and early cessations of rain fall in of Tigray in general and Abergalle in particular are common phenomena. And hence, studying different management options that could alleviate such problems and improve yield and quality of groundnut are important to groundnut production.

Thus, verifying and identification of the different integrated agronomic management aspects is of paramount importance to minimize the infection level and there by enhance production and productivity of groundnut.

1.2 Problem statement

The productivity of groundnut in Ethiopia is too low, estimated at 11.97qt/ha (CSA, 2010). The major factors that led to such low productivity are moisture stress, poor soil fertility, high temperature variation, erratic nature of rain fall, biotic factors (diseases and pests), lack of early maturing and high yielding varieties, and improper application of pre and post harvest management practices.

Aflatoxin, caused due to Aspergillus spp infection, is the most important limiting factor to groundnut production and utilization. Aspergillus is common and widespread in nature. It colonizes and contaminates grain before harvest and/or during storage. Groundnut is particularly susceptible to infection mainly following prolonged exposure to high temperature and stressful conditions. According to Dorner (2007), contamination of groundnut by alfatoxin is a worldwide problem that affects food safety.

Like in most groundnut producing areas of the world, groundnut production in Tigray is found to have high level of aflatoxin contamination. According to Abera (2008 un published data), groundnut samples collected from Tigray region had an average contamination of 29.56 ppb level which is higher than the acceptable limit for human consumption of 3 ppb (ICRISAT, 2009). Most countries have adopted regulation that limits the quantity of aflatoxin in food and feed to 20 ppb or less (Edlayn et al., 2008). Since use of improper management causes higher contamination resulting in lower quality, integrated application of relevant agronomic practices will help to reduce the contamination level and also improve productivity of the crop.

1.3 Research questions

The major research questions addressed in this research were, 'do integrated agronomic management practices enhance the productivity of groundnut and minimize the infection of Aspergillus spp? With this general question the study was employed to answer the following specific research questions:

What are the major pre harvest agronomic management practices that affect productivity of groundnut? Do agronomic management practices reduce Aspergillus infection?

1.4Justifications

The demand for high value and marketable crops is increasing through time. As a result, farmers with small land holdings are encouraged to produce high value crops including pulses. Groundnuts provide a vital source of cash income and nutritious, high protein food which could prevent child malnutrition. The low productivity of groundnut is influenced by moisture stress at the flowering to pod setting and hence use of supplementary irrigation and tied ridge is a way out to solve the moisture deficit. In addition, aflatoxin contamination is one of the major production constraints that influence groundnut quality. It is clear that production of groundnut is affected by a series of biotic and a biotic limiting factor.

Application of integrated agronomic management practices is crucially important so as to enhance groundnut production and reduce Aspergillus infection.

1.5 Objective

1.5.1 General objective:
To identify pre harvest agronomic management practices that improve groundnut kernel yield and reduce aflatoxin contamination due to Aspergillus invasion.

1.5.2 Specific Objectives:
- ✓ To determine the effect of soil amendments (inorganic fertilizer and gypsum) on groundnut grain productivity and level of infection due to Aspergillus.
- ✓ To determine the role of tied ridging on groundnut productivity and level of infection by Aspergillu.
- ✓ To determine the effect of supplementary irrigation in groundnut productivity and management of Aspergillus infection.
- ✓ To determine the combined effect of the different management practices on kernel yield and Aspergillus infection of groundnut

1.6 Hypothesis

Application of different pre-harvest practices (soil amendments and moisture improvement) and their combination result in a significant grain yield improvement and reduction in Aspergillus infection of groundnut.

Chapter Two: Literature Review

2.1 Taxonomy of Groundnut

Cultivated groundnut (*Arachis hypogae L.*) belongs to genus Arachis in sub tribe of Aeschynomeneae of family Leguminosae. Groundnut is self fertilized flower that withers soon after fertilization. Rates of natural out crossing are low and have been estimated between 0.7% and 2.5% pollinated crops (Wright, 2004).

Commercially cultivated varieties belong to variety hypogaea. Groundnut seed is a very delicate and is highly sensitive to different kinds of stress before, during, and after harvest including storage (Nigma et al., 2004). The cultivated peanut was probably first domesticated in the valley of Peru. It is annual herbaceous plant growing 30 to 50 cm tall. The leaves are opposite, pinnate with four leaflets (two opposite pairs; no terminal leaflet), each leaflet 1 to 7 cm long and 1 to 3 cm broad. The flowers are a typical pea flower in shape, 2 to 4 cm across, yellow with reddish veining. After pollination, the fruit develops into a legume 3 to 7 cm long, containing 1 to 4 seeds, which forces its way underground to mature. Hypogaea means under the earth.

2.2 Importance of Groundnut

Groundnut kernels are consumed directly as a raw, roasted or boiled or oil extracted from the kernel is used as culinary oil. The crop is also used as animal feed and industrial raw material. The multiple use of groundnut plant makes it an excellent cash crop for domestic markets as well as for foreign trade in several developing and developed countries. The oil content of the crop has been well appreciated and documented. Groundnut plays an important role in the oil economy of the world. It is the world's 4th most important source of edible oil (Bhatia et al.,

2006). Groundnut seed contains high quality edible oil (50%), easily digestible protein (25%) and 20% carbohydrates (Bhatia et al., 2006).

Groundnut is a multipurpose crop as a legume; it occupies an important position in cropping patterns based on cereal or root crop. Norman et al., (1996) has reported that the crop has substantial effect soil fertility, relative to non leguminous crops, which may be attributed to soil differences.

Groundnut is an important crop in its contribution to poverty reduction and food security. It is a basic food and cash crop and contributes to livestock feeding, export earnings and national trade development. Fana (2010) reported that, 63 derivatives could be produced from raw peanut products. For example, in Africa the leaves may be used for animal feed and cooked for human consumption. It can be boiled and broiled. It can be made into peanut oil, butter and flour. Study on groundnut also indicated that with the costs of animal protein becoming increasingly prohibitive, groundnut is becoming an even more important source of protein.

Peanuts are excellent source of nutrition for both human and animals. It was indicated that peanut kernel is rich in source of protein, oil, carbohydrate, vitamins and minerals (Wright, 2004). Groundnut is consumed raw, roasted, blanched, as peanut butter, crushed and mixed with traditional dishes as a sauce, a cooked paste. Findings of research indicated that Groundnut is an excellent source of oil for cooking and groundnut cake and haulms (straw, stems) are commonly used as animal feed (Okello et al., 2010).

It is well studied and documented that groundnuts thrive under low rainfall and as a legume it improves soil fertility by fixing nitrogen (Okello et al., 2010). And it is reported that the crop generally requires few inputs, making it appropriate for cultivation in low-input agriculture by smallholding farmers (Smartt, 1978). As a cash crop, it gives relatively high returns for limited land area and is well adapted to the hot, semi-arid conditions.

Groundnut as a legume, Rhizobium nodules on roots can provide a substantial portion of its nitrogen (N) requirement. Well- nodulated plants also tend to be more drought resistant and produce higher quality nuts (Weiss, 2000). Research study also indicated that rotation of groundnut with other crops have advantages, it reduces pests and disease damage (Brenneman et al., 1995).

2.3 Ecology and Groundnut production

Groundnut is a dicotyledonous plant which is widely distributed. The plant probably originated in the region of Brazil and has spread to many other countries especially tropical and subtropical countries of the world (Onwueme and Sinha, 1991). Groundnut like other annual oil seeds with a short growing season, it can adapt to a wide range of environment. However, some modern cultivars may exhibit a more highly developed local acclimatization. It is normally grown commercially below 1250 m, although specific varieties can be found at mach higher elevations (Weiss, 2000).

Groundnut is little affected by day length, being basically day neutral, but plants can be adversely affected by low light intensity during early growth and flowering. Bunch types are generally more severely affected by climatic variation within their normal range, the runner type least. While ambient temperature affects general plant growth and above 35 $^{\circ}$c adversely affects flower production, temperature in the soil zone in which pods develop, the geo carp sphere, has the greatest effect on yield. Temperature between 25 and 30 $^{\circ}$c appears to be the optimum, below 20 $^{\circ}$c retards development. Low temperature initially retard, then prevent growth, and the plant is generally killed by frost at any stage, while freshly dug and windrowed nuts that become frosted have reduced viability. However low temperature tolerance exists and could be further developed (Weiss, 2000).

Groundnut is essentially a tropical plant and requires a long and warm growing season. The favorable climate for groundnut is a well-distributed rainfall of at least 500 mm during the crop-growing season, and with abundance of sunshine and relatively warm temperature. Groundnut is drought tolerant, and to some extent it also tolerates flooding. A rainfall of 500 to 1000 mm will allow commercial production, although the crop can be produced on as little as 300 to 400 mm of rainfall. Groundnut thrives best in well-drained sandy loam soils, as light soil helps in easy penetration of pegs and their development and their harvesting (Weiss 2000).

Groundnut is grown on 26.4 million ha worldwide with a total production of 37.1 million metric ton and an average productivity of 1.4 metric ton ha^{-1} (FAO, 2003). Developing countries constitute 97% of the global area and 94% of the global production of this crop. The production of groundnut is concentrated in Asia and Africa 56% and 40% of the global area and 68% and 25% of the global production, respectively (Nigam et al ., 2004).

2.4 Effect of soil fertility amendment on yield of groundnut

Balanced fertilizer applications, based on soil tests provide adequate levels of phosphorus, potassium, calcium, sulfur, and magnesium. Nutrient availability depends on soil pH, organic matter content, and rate of release of nutrients from the soil minerals. The availability of other essential ions such as copper, boron, iron, manganese, and nitrogen may be low in alkaline soils (pH >8.5); while an acid soil (pH <6) may be deficient in molybdenum, manganese, sulfur, nitrogen, phosphorus, potassium, and calcium. Thus, depending on soil nutrient status and targeted yields and quality of groundnut production, soil test based application of nutrients is Paramount important.

2.4.1 The role of fertilizer (DAP) on yield of groundnut

Phosphorus is essential nutrient for crop growth and yield it promotes root growth, enhances nutrient and water use efficiency and increases yield. Due to the important role of phosphorus in physiology of the plant, application of phosphorus to deficient soils leads to increase in groundnut yield (Kamara et al., 2011).

The P requirement of groundnut is fairly high and seldom met from normal soil levels. Wide dispersal within the root zone is an important as the total amount, and thus higher application to preceding crops can improve the following groundnut yield. The requirement p fertilizer in groundnut production was well studied. Phosphorus fertilizer rates exhibited significant effect on number of pods and seeds/plant, weight of pods and seeds/plant, 100 seed weight as well as seed oil and protein (Mirvat et al., 2006).

The addition of phosphorus-containing fertilizer to peanuts is generally not needed if it is applied to other crops in the rotation. As a result soil testing is the best option for the application of phosphorus during peanut production (David, 2011). However, other study on P response of groundnut explained that there is a positive relation of kernel yield of groundnut with p application. According to Mupangwa and Tagwira (2005), Phosphorus significantlly (p <0.001) increased kernel yield across the seasons. Application of 8.5, 17 and 34 kg ha^{-1} P increased kernel yield by 39, 40 and 51% over the zero P treatment. Besides, study of groundnut on different soil types showed that there exists a significant yield response on low available P plant nutrients. Study work showed that 8.5 kg P ha^{-1} gives the highest rate of return from groundnut grown in acid sandy soils (Chikowo, 1998). Lack of kernel yield responses to P at a given site could have been a result of other soil limiting factors (Chikowo, 1998). Field and greenhouse studies by Otani and Ae (1996) showed that groundnut took up more P from low P soils than other crops such as sorghum. Ae et al. (1996) also attributed lack of yield response to P by groundnut to the exceptional ability of the legume to extract P from low P soils. Ae et al. (1996) also reported that groundnut releases relatively high exudates which solubilize Al-bound P (Mupangwa and Tagwira, 2005).

With regard to potential response of groundnut to P fertilizer and in comparison to locally available P sources i.e., single superphosphate (SSP) and fused magnesium phosphate (FMP), and farmyard manure (FYM); there was a large response to P, applied as SSP. Pod yield was significantly correlated with P fertilizer application. Groundnut showed a strong response to P fertilizer regardless of source. Depending on the crop year, application of fertilizer P alone increased pod yield by 26 to 66% and seed yield by 24 to 102% Application of SSP along with 500 kg lime/ha gave a higher pod and seed yield than application of FMP and lime. It is reasoned that the higher pH of the fused magnesium phosphate (FMP) source, which is already rich in Ca oxide (CaO) and Mg oxide (MgO), may be a less plant available P source, especially when it is combined with 500 kg lime/ha (Nguyen,2003).

Nitrogen is an essential element and important determinants of plant development. Groundnut is a legume, and Rhizobium nodules on roots can provide a substantial portion of its nitrogen requirement. An additional of nitrogen as a fertilizer is not largely used for peanuts. Even though symbiotic relationship provides sufficient nitrogen for peanut production if the roots are properly nodulated. Peanut seed or fields had to be ensured that there must be adequate levels of Rhizobia in each field. Peanuts generally react better when the previous crop is fertilized than to a direct addition of fertilizer, yet direct fertilization can be an advantage in the case of the light, sandy soils. Well nodulated plants also tend to be more drought resistant and produce higher quality nuts (Weiss, 2000). The nitrogen requirement was well studied. Addition of N fertilizer generally increases root-shoot ratio (Kumar and. Rao. 1991). Yield advantage of N and P fertilizers application was reported in groundnut. As the soil was originally poor in N and P pod yield responses were high. Study of NP application on groundnut indicated good groundnut yield from application of 40-60 kg NP per hectare can be harvested. It resulted in greater utilization of assimilates into the pods and ultimately increased number of mature pods per plant, 100-seed weight and yield (Hossain et al., 2007).

2.4.2 The role of Calcium on yield of groundnut

Maximum use of calcium (Ca) by plants occurs from the appearance of pegs until pods are mature, while high level of calcium may reduce the incidence of pod and root rot. On some low calcium –content soils, gypsum can also increase seed –oil content. Use of gypsum as a calcium fertilizer for peanut is well known and adequate quantity of calcium during pegging around the pegging zone helps for the proper development of disease free peanut production (Liming and Warren, 2011).

Study on the requirement of calcium in peanut production indicated that calcium is the most critical nutrient for achieving high yields and grades. Low level of calcium causes several serious production problems, including unfilled pods (pops), darkened plumules in the seed and poor germination (Faujdar and Oswalt, 1995).

It is reported that presence of enough calcium content in the soil leads to prevent of black hallow and cracked pods, decrease of aflatoxin production and consequently decrease decayed pod of peanut. Besides, calcium is one of the most important nutritional elements to gain high yield and quality of peanut (Alireza et al., 2012). Depending on the soil test, 300-500 kg ha -1 of gypsum should be applied at a depth of 3-5 cm prior to pegging (Faujdar and Oswalt, 1995).

The response to Ca applied as calcitic lime and gypsum was variable. Adequate Ca supply in the podding zone is critical for the production of quality kernels. The Ca requirement for kernel development is taken up directly by the pod from the soil (Zharare et al., 1993; Zharare, 1996). Pods are poor absorbers of Ca and hence require that the soil has ample Ca levels.

 Desai et al. (1999) reported that on acid soils, top dressed gypsum at flowering and pre plant broadcast lime gave similar groundnut yield responses. Smith (1995) also reported that lime is a more suitable source of Ca than gypsum for groundnut grown on acid light textured soils because of the slow release of Ca. Gypsum is subjected to leaching and this could be one of the reasons why in significant yield responses were observed in groundnut production.

Soil moisture is critical to the availability of Ca to groundnut. Soil moisture status has a significant role during application of calcium. This could have improved the solubility of calcitic lime and hence the supply of Ca from this source. Smith (1995) reported that soil moisture is very critical in the pegging zone during peak periods of Ca uptake by pods. Groundnut growing in dry topsoil but with roots in moist subsoil show poor pod development and kernel abortion because Ca absorbed by the roots is not channeled to the developing pod (Faujdar and Oswalt, 1995).

2.5 Effect of moisture conservation practices on yield of groundnut

2.5.1 The role of Tied ridge in moisture conservation

Contour ridges (or contour furrows) involve making ridges along the contour at a spacing usually of some 1-2m.In Kenya and Tanzania, ridging is normally done for crops such as potatoes, tobacco, groundnuts and even for maize (Assmo and Eriksson, 1999).

Tied ridging is a conservation tillage system consisting of ridges which are cross-tied at regular intervals with small dams. The basins formed between the ties prevent the water from flowing off the field. Tied ridging increases yields particularly when combined with improved soil fertility. Less tillage work required in following season reduces soil erosion and conserves moisture. Ridging systems are mostly suited for areas with an annual rainfall ranging from 350 to 750 mm (Critchley and Siegert 1991). In the semi-arid areas, tied ridges are made by modifying normal ridges. The technique involves digging major ridges that run across the predominant slope, and then creating smaller transverse sub-ridges (or cross-ties) within the main furrows. The final effect is a series of small micro-basins that store rainwater in-situ, enhancing infiltration. Depending on the system, the crop is planted at the side of the main ridge, to be as close as possible to the harvested water while also avoiding water logging in case of prolonged rains.

In line with this, studies on tied ridges have been found to be very efficient in storing rain water. This has resulted in substantial grain yield increase in some of the major dry land crops such as sorghum, maize, wheat, and mung beans in Ethiopia (Georgis and Takele, 2000). The average grain yield increase (under tied ridges) ranged from 50 to over 100 percent when compared with the traditional practice. This increase, however, will vary according to the soil type, slope, rainfall and the crop grown in dryland areas such as Kobbo, Nazreth, Meiso, Mekelle and Babilie of Ethiopia (Mati, 2005).

Study on moisture stress mitigation was aimed at determining appropriate strategy (ies) that would best conserve soil and moisture on farmlands to result in optimal grain and Stover yields and improve on farmers' livelihoods. Low soil organic carbon content, and low inherent soil fertility are among major factors responsible for low crop yields (Das *et al.*, 1991). Surface configuration, such as tied ridges; have been used to trap runoff when rainfall exceeds infiltration (Hulugalle, 1990) in drought-prone shallow soils of the West African Sahel. Ridges are advantageous on some nutrient deficient soils in Savannah region of Nigeria to concentrate the fertile top soil and to conserve water (Lal, 1995).

In the rain fed Alf sols and associated soils of semi arid tropics, utilization of tied ridge and locally available organic manures such as FYM in combination with inorganic N and P fertilizers can result in alleviation of climatic and soil-related constraints like low soil water retention and low soil fertility. Study on Alf sol indicated that soil water storage and soil nutrient management, as influenced by land configuration and manures were the critical factors to increase the yield of sorghum. In relation to either open ridge or planting on the flat bed, tied ridge increases soil water content and grain and straw yield of sorghum irrespective of rainfall category (Selvaraju et al., 1999).

2.5.2 The Role of Supplementary Irrigation in dryland farming

The drylands of Ethiopia comprise about 70 percent of the total landmass and 45 percent of the arable land, including arid, dry semi-arid, moist semi-arid and parts of the sub-moist zone.

However, these areas contribute only 10 percent of the total crop production (Mati, 2005). About half of the arable land is in the arid and semi-arid regions, and most rural people living in such areas depend on small-scale dryland agriculture.

The drylands are characterized by a severely fragile natural resource base. Soils are often coarse-textured, sandy, and inherently low in organic matter and water-holding capacity, thereby making them easily susceptible to both wind and water erosion (Mati, 2005). As a result, crops can suffer from moisture stress. Farm productivity has declined substantially and farmers have found themselves sliding into poverty (Kidane, 1999).

Adequate water available throughout the peanut life cycle is important for optimal plant growth and development. Drought or flood can have tremendously negative impacts on peanut yields and quality. Likewise, pest infestation and severity of damage from these pests is influenced by available water, either in the form of rainfall or irrigation. Irrigation minimizes risk and enhances consistency of yield. Low and erratic rainfall, poor or steeply sloped soils and a short cropping season make the uncertainty of dryland agriculture in semi-arid regions. Low soil fertility and drought are the main factors affecting the productivity and sustainability of rain fed agriculture. High inter annual variability and erratic rainfall distribution in space and time result in water-limiting conditions during the cropping season.

Crop production under dry farming depends on water supply from current rainfall and water stored in the root zone of the soil at sowing.

Due to the dry conditions and frequency of drought events, the role of supplemental irrigation at late growing season crops is highly emphasized to avoid complete failure of crops and to minimize negative impacts on yield. The positive effect of supplemental irrigation on response to

N application suggests that soil moisture is the chief constraint to crop growth and N may be the second most-limiting factor in this semi–arid region (Sene and Badiane, 2005). Study on groundnut production under rain fed and irrigation condition indicated that higher yield of groundnut was obtained at irrigated condition. Accordingly, at irrigated and rain fed groundnut production 5827 and 3482 kg/ha of kernel yield, respectively was obtained (Getinet et al., 1997).

2.6 Importance of Aspergillus spp in groundnut production and utilization

Groundnuts are prone to infestation by two closely related fungal species, *Aspergillus flavus* and *A. parasiticus*. Both fungi species produce a highly toxic group of mycotoxins known as aflatoxins. Health effect on human and livestock due to consumption of aflatoxin contaminated foods include impaired growth, liver and other cancers, immune-suppression, synergisms and death. Aflatoxins are natural toxic chemical substances produced by *Aspergillus flavus* and *A. parasiticus*. These toxins can contaminate any array of crops including maize, groundnuts and tree nuts. It was reported that in April 2004, high aflatoxine outbreak occurred in rural Kenya, resulting in 317 cases and 125 death; aflatoxin contaminated home grown maize was the source of the outbreak (Lewis et al., 2005). Besides, study on aflatoxin in West Africa indicated that many individuals were malnourished and chronically exposed to high levels of aflatoxin through their diets (Gong et al., 2002).

Due to deleterious health hazards, aflatoxin contamination significantly restricts the volume of groundnut exports from sub-Saharan Africa (Freeman et al., 1999). International trade restriction is particularly serious, because of the European Union's (EU) imposition of a new aflatoxin regulation, which is stricter than that was suggested by the Codex Alimentarius Commission (Ntare et al., 2005). The potential seriousness of the export-restricting effect of aflatoxin contamination in the groundnut sectors in many African countries has been well documented (Otsuki et al., 2001).

The impact analysis of the European Union's new harmonized aflatoxin limits on exports from Africa indicated that 1 percent lower maximum allowable level of aflatoxin contamination will decrease groundnut trade by 1.3 percent. The results of the study suggest that the implementation of the new (and more stringent) EU aflatoxin regulations will impact adversely on African exports of even cereals, dried fruits, and nuts to Europe. More specifically, the study suggests that even though the new EU standard would decrease health risk by roughly 1.4 deaths per billion a year, it will result in a \$670 million (or 64 percent) reduction in African exports, in contrast to a regulation based on an international standard suggested by Codex guidelines (Waliyar et al., 2007).

2.7 Biology of Aspergillus spp

2.7.1 Over view of Aspergillus spp
Aspergillus had become one of the best-known and most studied mould groups. Their prevalence in the natural environment, their ease of cultivation on laboratory media and the economic importance of several of its species ensured that many mycologists and industrial microbiologists were attracted to their study. Aspergilli grow abundantly as saprophytes on decaying vegetation where they are found in large numbers from moldy hay, organic compost piles, leaf litter and the like. Furthermore, this common mould is involved in many industrial processes including enzymes (e.g. amylases), commodity chemicals (e.g. citric acid) and food stuffs (e.g. soy sauce). Several species contaminate grains and other foods with toxic metabolites that are a threat to the health of humans and other animals. Certain Aspergillus species also can infect humans directly

causing both localized and systemic infections, especially in immune compromised individuals (Bennett, 2010).

2.7.2 Ecology of Aspergillus

Microorganisms normally require certain amount of moisture in order to grow and multiply. Apart from moisture content of the seed, several environmental factors influence the extent to which fungal growth and aflatoxin contamination occurs.

Study indicated that when the seed moisture exceeds 9% at equilibrium humidity of 80% and temperature of 30 °c the chance of invasion by *Aspergillus flavus* increase drastically (Diener and Cole (1982). Aspergillus spores are common components of aerosols where they drift on air currents, dispersing themselves both short and long distances depending on environmental conditions. One of the defining characteristics of the entire fungal kingdom is its distinctive nutritional strategy.

Species of Aspergillus are typical examples of the fungal life style. They are most often found in terrestrial habitats and are commonly isolated from soil and associated plant litter. In the ecosystem, different substrates are attacked at different rates by consortia of organisms from different kingdoms (Bennett, 2010).

2.7.3 Identification and characteristics of Aspergillus spp

The defining characteristic of the genus Aspergillus is the aspergillum-like spore-bearing structure. The size and arrangement of the conidial heads as well as the color of the spores they bear are important identifying characteristics. *A. niger* group bear black spores, the *A. ochraceus* group is yellow to brown, while *A. fumigatus*, *A. nidulans*, and *A. flavus* are green (Bennett, 2010).The major cultural features used in species identification are the colour of the colony, the growth rate and thermo tolerance. Aspergilli have varying morphological and growth response to different nutrients so it is important to standardize conditions. Species identification depends upon pure cultures grown on known media. In addition to the conidiophore, other morphological structures useful for identification include cleistothecia, Hulle cells and sclerotia. Both cleistothecia and sclerotia are closed and usually round structures about the size of a poppy seed that may be so abundant as to dominate a colony. Cleistothecia are the sexual reproductive stage and contain the meiotic ascospores borne within asci (Bennett, 2010).

2.8 Aflatoxins and its Effect on Disease

Aflatoxins are one of the major mycotoxins of agricultural importance. Mycotoxins are secondary metabolites of fungi. The major fungal genera producing mycotoxins include Aspergillus, Fusarium and Penicillium. Many foods and feeds can become contaminated with mycotoxins when invaded by such fungi. Mycotoxins can form in commodities before harvest, during the time between harvesting and drying, and in storage. Commodities and products frequently contaminated with mycotoxins include corn, wheat, barley, rice, oats, nuts, milk, cheese, peanuts and cottonseed. Mycotoxins produce a wide range of adverse and toxic effects in animals in addition to being food borne hazards to humans (Michael et al., 2006).

Aflatoxins are still recognized as the most important mycotoxins. They are synthesized by only a few Aspergillus species of which *A. flavus* and *A. parasiticus* are the most problematic.

The expression of aflatoxin - related toxicoses is influenced by factors such as age, nutrition, sex, species and the possibility of concurrent exposure to other toxins. The main target organ in mammals is the liver so aflatoxicosis is primarily a hepatic disease. Conditions increasing the likelihood of aflatoxicosis in humans include limited availability of food, environmental conditions that favor mould growth on foodstuffs, and lack of regulatory systems for aflatoxin monitoring and control (Henry et al., 1999; Williams et al., 2004). *Aspergillus flavus* and

A. parasiticus are moulds that grow on a large number of substrates, particularly under high moisture conditions (Bennett, 2010).

The staple commodities regularly contaminated with aflatoxins include cassava, chillies, corn, cotton seed, millet, peanuts, rice, sorghum, sunflower seeds, tree nuts, wheat, and a variety of spices intended for human or animal food use (Weidenborner, 2001). Human exposure to aflatoxins exposure is difficult to avoid because *A. flavus* grows aggressively in many foods at all stages of the food chain in the field, in storage and in the home (deVries et al., 2002). Evidence for acute human aflatoxicosis has been reported from several underdeveloped countries such as India and Thailand (Bennett, 2010).

2.9 The role of Agronomic Practices in minimizing Aspergillus infection

The main factors leading to aflatoxin contamination includes use of damaged and loose shelled kernel as seeds delayed harvesting after physiological maturity, high moisture content in pods, inadequate protection from rain, pests and diseases. Aflatoxin management should start in farmers field with proper crop production management and handling, (UNIDO, 2010).

The development of *A. flavus* and aflatoxin contamination are more serious in field where a field is poorly managed. Contamination takes place in production of edible peanuts, including in the field in late season condition of drought, during transport from field to sale. Pre harvest Aflatoxin and cyclopiazonic acid contamination of peanut is a great economic problem for the industry and a danger to human and animal health.

Severe drought stress on peanuts late in the growing season mostly resulted in decrease in harvest, besides low quality yield, higher incidence of *A.flavus* group invasion and aflatoxin production. Rachaptie et al (2002) observed that aflatoxine contamination was wide spread with severe and prolonged end of season drought associated with elevated soil temperature. Other study indicated that aflatoxin contamination of groundnut is a major problem in most of the groundnut-producing regions across the world. The occurrence of drought during the late seed-filling period is a key contributing factor. It is caused by the growth of the moulds *Aspergillus flavus* and/or *Aspergillus parasiticus* (Abdalla et al., 2005).

Aflatoxins occurring naturally in foods and feeds may be reduced by a variety of procedures.

Literature review on improved farm management practices, early planting and on time harvesting as well as soil fertility amendments practices explained that these practices has an effect on reducing aspergillus infection on groundnut. Tied ridge as one of agronomic practice in conserving moisture can have a direct linkage in reducing Aspergillus infection. In relation to this study indicated that tied ridging, particularly when combined with improved soil fertility less tillage work required in following season reduces soil erosion and conserves moisture. Ridging systems are mostly suited for areas with an annual rainfall ranging from 350 to 750 mm (Critchley and Siegert, 1991).

Management practices implemented at end-of-season which have been developed to improve water retention after cessation of rains had a significant effect in minimizing aflatoxin contamination (Craufurd *et al.,* 2006).

Aflatoxins are not visible neither do they have a particular flavor. Crop husbandry practices, climatic conditions, and soil factors in addition to host plant susceptibility, significantly influence aflatoxin contamination.

Literature review expressed that groundnut is one of the most susceptible crops to aflatoxin contamination (Okello et al., 2010) yet it is a major staple, widely cultivated and consumed. A study on maize research indicated that management practices that improve plant health strongly discourage aflatoxin development. Timely planting, adequate fertility, good weed and insect control, supplemental irrigation, suitable plant population, and hybrid selection help in reducing aflatoxin potential (Erick, 2009).

Chapter three: Materials and Methods

3.1 Description of the study area

The study was conducted during the 2011 cropping season at Tanqua Abergelle woreda at *Tabia* Lemlem and Hadinet in the central administrative Zone of Tigray, northern Ethiopia on sandy soil. Tanqua Abergelle is located 120 kms far from Mekelle at an altitude of 1500 m.a.s.l , 13^0 14' 06"N Latitude & $38^0 58'50$" E longitudes (Fig. 1). It is agro-ecologically characterized as hot warm sub-moist low land (SMl-4b). The mean annual rainfall and temperature range of the wereda is 350 – 700mm & 24-41^{0}c, respectively (Legesse, 1999). The mean maximum and minimum temperature of the wereda during the 2011 growing season was 36.4^{o}c and 19.5^{o}c, respectively. While the total rain fall of Hadinet and Lemlem experimental sites during the 2011 growing season was 277mm and 314 mm respectively.

Figure 3.1 Map of Tigray and study area

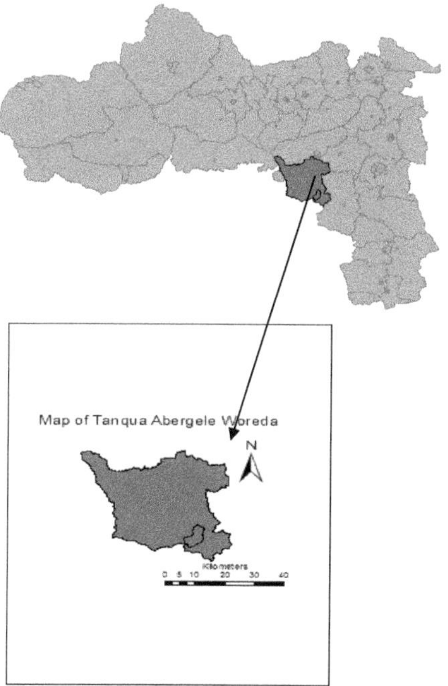

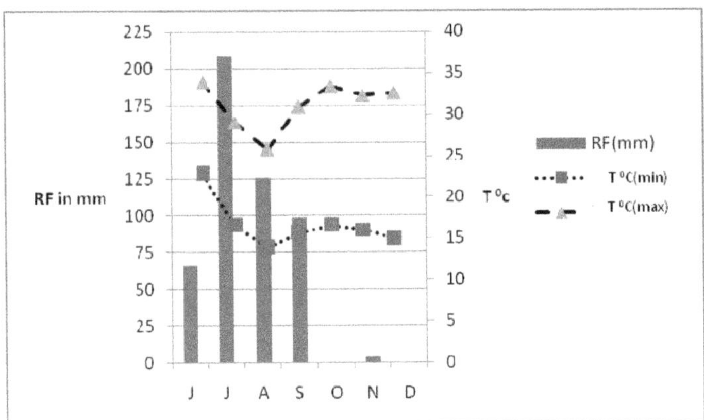

Figure 3. 1 Mean monthly rain fall (mm) and maximum and minimum temperature of Abergelle Wereda during the 2011 growing season.

3.2 Treatments and Experimental Design

The groundnut cultivar Sedi which is popular and widely used in the study area was used in the experiment. The experimental plot was arranged in RCBD with three replications. Different farmers field were considered as a replication. The total plot size used was $9m^2$ (3m X 3m), having 60cm distance between rows and 20cm between plants. The net harvestable area was $5.4m^2$(1.8 m x 3 m) leaving one outer most row as border.

The experiment was conducted to determine the effect of Integrated On–farm Agronomic Practices on the productivity of groundnut and development of Aspergillus infection on groundnut kernel.

Table 3.1 list of treatments and treatments combinations

Number	Treatments
1	C
2	GP
3	SI
4	TD
5	F
6	GP + SI
7	GP + TD
8	GP + F
9	SI + TD
10	SI + F
11	TD + F
12	GP + SI + TD
13	GP + SI + F
14	GP + TD + F
15	SI + TD + F
16	SI + TD + F + GP

C = Control, GP = gypsum, SI = supplementary irrigation, TD = tied ridge, F = fertilizer (DAP)

DAP (100 kg/ha), as a source of Phosphorous was applied at planting, application of gypsum 672kg/ha was performed at flowering stage; while Tied ridge and application of supplementary irrigation were done at the end of rainy season. The tied ridge had 20 X 30cm depth of the soil.

3.3 Crop water requirement and irrigation scheduling

Meteorological data (maximum and minimum temperature, relative humidity, sunshine hours, rainfall and wind speed) were collected from National Meteorology Agency Mekelle branch. Crop water requirement was determined using modified FAO Penman Menteith method (Allen et al., 1998). Reference evapotanspiration (ET0) was calculated based on the climatic data using CWAT software. While the other parameters were calculated as:

Crop water requirement (Etc) = Eto * Kc (crop coefficient at each growth stage of the crop)

Gross irrigation (GI) =Net irrigation (NI)/application efficiency (Ea), Ea was taken as 85%

NI = Etc – Pe(Pe =effective rain fall), but in my case since there was no rain fall at supplementary irrigation time   NI = Etc. Irrigation was started at late September for both locations. Accordingly, irrigation water was applied for the specific treatments/plots every five days as supplementary as indicated in (Table 3.2).

Table 3.2 Irrigation scheduling of ground nut as a supplementary irrigation

location	Date	days	Crop growth stage	RF(mm)	Net irr(mm	Gr.irr mm	Volume per plot(mm) (VP)	Number of bucket (VP/20 liter)
Hadinet	23/9/2011	81	END	00	15.75	18.5	166.5	8.3
	28/9/2011	86	END	00	15.1	17.8	160.2	8.0
	3/10/2011	91	END	00	14.4	16.9	152.1	7.6
	8/10/2011	96	END	00	13.4	15.7	141.3	7.1
Lemlem	25/9/2011	83	END	00	15.75	18.5	166.5	8.3
	30/9/2011	88	END	00	15.1	17.8	160.2	8.0
	5/10/2011	93	END	00	14.4	16.9	152.1	7.6
	10/10/2011	98	END	00	13.4	15.7	141.3	7.1

3.4 Field data collection

3.4.1 Phenology

Data on days to 50% emergence, 50% flowering, 50% pod setting and 90% physiological maturity were recorded when the plants per plot have reached their respective phenological stages.

3.4.2 Yield and yield components

Yield components including number of pods/plant and number of seeds /pod were determined from measurements of five randomly selected plants per net plot area. Pods were shelled and shelling percentage per plot was computed. Harvest index was also calculated. Dry pod and seed yield per plot area were recorded and converted to yield per hectare. Hundred seed weight was also measured.

3.5 Soil sample Collection and Analysis

At planting, soil samples were randomly collected from each field to a depth of 0-40cm with auger. Composite soil samples were prepared from the collected samples. The samples were oven dried at 105 $^{\circ}$c for 24 hours and ground to pass through a 2-mm sieve for physical and chemical analysis. Soil samples were analyzed at TARI soil laboratory for relevant soil parameters (bulk density, permanent wilting point (PWP), field capacity (FC), organic matter, pH, Nitrogen, Phosphorus, Potassium, exchangeable Ca, Mg and soil texture).

Samples were also collected from each plot at harvest and the moisture content was determined by oven drying at 105 $^{\circ}$C till constant weight. The following procedures were used to determine soil parameters:

In the laboratory exchangeable bases (potassium, calcium and magnesium) were determined by ammonium acetates method. Besides soil texture was determined by Hydrometer method

Total nitrogen: was determined by using Kjeldahl method as described by Jackson (1967).

Available phosphorus: using spectrophotometer following the Olson extraction method described by Olson and Dean (1965).

pH: measured from the composite soil sample in a suspension of a 1: 2.5 soil to water ratio as described by Jackson(1958).

Organic matter: Was determined by walkley and Black wet oxidation organic Carbone method as described by Jackson (1967).

3.6 Isolation and identification of Aspergillus in Laboratory

Groundnut kernel seed was examined in laboratory and the incidence and severity of Aspergillus infection was recorded based incidence and severity percentage as described by Goanzalel et al (1995). Hundred kernels of groundnut samples from bulk were taken randomly from each experimental plot for lab analysis. Using the blotter plate method (ICRISAT, 2005) the seeds were soaked in distilled water for three minutes and surface sterilized by soaking in 1% sodium hypochlorite solution for one minute and rinsed three times with distilled water so as to protect from other external contamination. Then, the kernels were subjected to examination by placing them on moistened germination paper in the sterile plastic tray at an equal distance sealed with paraflin and were grown at 12 hours dark and light cycle , incubated at dark room for 5 days at 28°c with florescence light. Then kernel infected by Aspergillus spp (*A.fluvs* and *A.nigr*)were counted based on the color of the fungus (Pit et al., 1983).

 Incidence and severity of each fungus types were determined according to Gonzalel et al (1995). Incidence as the percentage of number of Aspergillus infected samples by the fungus to the whole sample size as:

I = n/N *100, n represents number of Aspergillus **infected sample** and N represent the total **sample size**

The severity (%) of each fungus was expressed as a percentage of infected grains per 100 plated seeds as:

S (%) =n/N *100,

Where n is the number of **infected seeds** and N is the **total plated seeds**

3.7 Statistical Analysis

Analysis of variance (ANOVA) was carried out on yield and yield components of groundnut using Genstat software 13th editions. Treatment means showing significant differences at 5% level of significance were compared using Duncan's Multiple Range Test (DMRT). As the value consisted of small whole numbers and range between 0 and 30%   square root transformation method was employed before statistical analysis. Percentage data on fungal infection were transformed using square root method as described by Gomez and Gomez (1984).

Chapter Four: Results and Discussion

4.1 Soil characteristics

Results of soil analysis for soil sample taken before planting indicated that the proportion of sand, silt and clay was 94%, 1% and 5%, respectively. As a result the textural classes of the experimental soils are classed as sandy (Table4.1).The available phosphorus content and organic carbon was low in both sites (< 10 mg /kg and < 2%, respectively), however Lemlem site has lower Phosphorus than Hadinet (Table4.1). The pH, values of soil in Hadinet and Lemlem, was 7.04 and 7.06 respectively which indicated that the experimental soils are almost neutral and within the ideal pH range value for groundnut production. In addition the EC of Hadinet and Lemlem were 0.04 and 0.05mmhos/cm, respectively indicating that the sites were non saline.

Table 4.1 Soil characteristics of experimental site

Soil characteristics	Location	
	Hadinet	Lemelem
pH $_{(1:2.5)}$	7.04	7.06
EC $_{(1:2.5)}$	0.04	0.05
OC (%)	0.26	0.14
Olsen-P(mg /kg)	6.41	3.51
Total - N (%)	0.08	0.05
Exch Na (mg/kg)	31.33	14.33
Exch K (mg/kg)	21.33	10.67
Exch Ca + mg (mlieq/lit)	6.47	6.73
Exch Ca (mlieq/lit)	1.53	1.4
Sand (%)	94	94
Silt (%)	1	1
Clay %)	5	5
texture	sand	sand

Table 4.2 Field capacity (FC), permanent wilting point (PWP), and bulk density of experimental site

Location	Root depth(cm)	Texture	Bulk density	FC	PWP	TAW (%) FC -PWP	TAW(volume),cm	TAW mm/m
Hadinet	0 - 40	sand	1.12	6.10	1.25	4.90	5.51	55.1
	41 -80	sand	1.22	9.60	4.28	5.35	6.53	65.31
Lemelem	0 - 40	sand	1.11	6.12	1.25	4.87	5.12	54.22
	41 -80	sand	1.23	9.64	4.28	5.36	6.57	65.75

TAW=total available water

Total available water for the experimental areas was 55.1mm and 54.22mm at 0-40 cm root depth at Hadinet and Lemlem, respectively. While the total available water at a depth of 41-80cm was 65.31 and 65.75mm/m at Hadinet and Lemlem experimental sites, respectively (Table 4.2).This result also agrees with the results of Allen et al.(1998) where he describes the values of total available soil water of sandy soil ranges between 50mm – 110mm/m

4.2 Soil water content at harvest

Soil moisture content was measured at harvest to make comparison among different management options in retaining water so that their effect on improving yield of groundnut and thereby impact in minimizing Aspergillus infection could be estimated. Analysis of variance on soil moisture determination at harvesting in Hadinet indicated that no significant difference ($P < 5\%$) was observed among the different management options at a soil depth of (0 – 40 cm). But significant difference (at $P < 5\%$) was observed at soil depth of 41- 80cm (Table4.3).The integrated management options showed an effect on moisture content percentage at both soil depth in lemlem experimental site. The highest moisture content (**9.04%**) was recorded in Hadinet in (41 – 80 cm) soil depth, where gypsum + **supplementary irrigation + tied ridging** management options were practiced while the lowest moisture content at control (Table4.3). In Lemlem experimental site the highest (13.68%) and the lowest (3.93 %) moisture content were recorded in supplementary irrigation + tied ridging management options and control at soil depth of 0 – 40cm, respectively this could be due to the better rain fall in amount and distribution than Hadinet. While in a soil depth of 41 - 80cm 14.63 percent moisture content was recorded in gypsum + supplementary irrigation + tied ridging and fertilizer combination management options were applied (Table4.3).

Table4.3 Effect of Agronomic management options on moisture retention in different soil depth at harvesting

Treatments Mgt options)	Moisture content (%)			
	Hadinet		Lemlem	
	Soil depth (cm)		Soil depth(cm)	
	0 – 40	41-80	0 – 40	41-80
C	1.87	3.06a	3.93a	4.32a
F	3.57	5.63ab	7.30b	8.85bc
GP	4.16	6.05b	7.37b	8.35ab
GP + F	3.12	8.86cd	10.09cde	9.73bc
GP + SI +TD	5.15	9.04d	12.61ef	14.23de
GP +SI	5.07	7.35bcd	8.50bc	9.86bc
GP +SI + F	4.23	6.43bcd	9.62bcd	9.74bc
GP +TD	3.22	6.13b	8.02bc	8.94bc
GP +TD + F	3.78	6.19b	9.20bc	10.16bcd
SI	5.51	7.55bcd	11.93def	12.22bcde
SI + F	4.32	7.70bcd	10.15cde	11.58bcde
SI + TD	6.46	6.66bcd	13.68f	14.31de
SI +TD + F	5.60	8.92d	11.96def	13.10cde
SI +TD + F + GP	4.34	7.31bcd	12.91f	14.63e
TD	4.03	6.90bcd	7.04b	8.30ab
TD + F	3.39	5.29ab	9.07bc	10.87bcde
DMRT[a] 0.5	ns	2.718	2.690	4.267
CV %	35.9	23.9	16.8	24.2

[a] Any means having the common letter are not significantly different at the 5% level of significance

4.3 Effect of agronomic practices on crop parameters

4.3.1 Crop phonology

The effect of the different agronomic management options tested is presented in Table 4.4. The analysis of variances indicated that no significant difference (P <5%) were observed among the different management practices in days to 50% flowering. However, the duration to full maturity day was significantly longer in one of experimental sites (Lemlem) and this could be due to the better rain fall amount and distribution that created a difference among the treatments. Early maturity (97 days) but lower yield was recorded on the control, but this earliness might be forced maturity that resulting from stress. On the other hand, the longest maturity days 109) were recorded in SI + TD + F management options implemented (Table 4.4).

Table4.4 Effect of agronomic practices on phenology and plant height of groundnut

Treatments	Hadinet			Lemlem		
	DTF	DTM	PH (cm)	DTF	DTM	Ht (cm)
C	39.33	107.00	8.77a	39.67	97.67a	14.40
F	41.33	108.00	17.47b	41.33	105.00bc	18.53
GP	39.33	108.67	19.87 bcd	39.00	102.67ab	20.00
GP + F	44.00	111.00	18.20bc	41.33	105.33bc	18.87
GP + SI +TD	39.33	109.00	21.73 bcde	40.67	107.67bc	18.93
GP +SI	39.33	112.67	22.43 bc	40.67	107.33bc	19.67
GP +SI + F	41.33	110.00	23.77 de	40.33	107.67bc	17.93
GP +TD	40.33	110.67	18.60 bc	39.00	104.33bc	20.07
GP +TD + F	40.33	110.00	20.73 bcde	40.00	108.00bc	18.53
SI	39.33	106.00	20.53 bcde	39.00	108.33c	19.73
SI + F	41.33	111.00	21.73 bcde	43.00	108.00bc	17.60
SI + TD	40.33	110.00	25.13 e	40.67	107.33bc	19.53
SI +TD + F	40.33	110.67	23.87 de	41.33	109.33c	17.93
SI +TD + F + GP	41.33	112.00	24.60e	40.33	106.67bc	20.07
TD	39.33	107.00	17.83bc	39.00	104.00bc	20.53
TD + F	40.33	110.00	19.13bc	42.00	104.67bc	21.53
DMRT [a] 0.5	ns	ns	4.621	ns	5.560	Ns
CV %	5.32	2.81	13.7	6.63	4.6	11.4

C=Control, F=Fertilizer (DAP), GP=Gypsums, SI=Supplementary Irrigation, TD =Tied ridge
DTF=days to flower, DTM=days to maturity, PH= plant height
[a] Any means having the common letter are not significantly different at the 5% level of significance

Mean while, significant difference (p< 0.05) was observed in plant height at Hadinet. Application of DAP fertilizer, tied ridge, supplementary irrigation and gypsum showed a significant effect (P<5%) over the control. No consistent result was observed through the different combinations, and hence no significant results were observed among most of the treatment combinations (Table 4.4). Over all combined applications showed a significant effect on plant height over the control. The maximum t height (25.13cm) was obtained from combined treatment of supplementary irrigation and tied ridging. While the shortest height (8.77cm) was observed for the control at Hadinet (Table 4.4).This variation in the significance across location might have occurred as a result of moisture difference during the growth period.

4.3.2 Yield components

4.3.2.1 Number of pods per plant

The analysis of number of pods per plant indicated that there was significant (P <0.001) difference among the evaluated management options at both locations. Highest (23.53 and 21.8 pods/plant) were recorded in supplementary irrigation + tied ridge + fertilizer (DAP) and gypsum + supplementary irrigation management option in Hadinet and Lemlem, respectively (Table 4.5).While the lowest (5.97 and 8.33 pods/plant) were recorded in Hadinet and Lemlem, respectively, in the control treatment.

Application of SI, TD, F and GP differed significantly on number of pods per plan over the control when these treatments were practiced alone (table 4.5). From all combination practices application of SI + TD + F and SI TD + GP showed a Significant difference on number of pods per plant (P <5%) over the control as well as other combinations (Table 4.5). from this result it can be suggested that application of supplementary irrigation and tied together with fertilizer application showed better number of pods per plant. And hence ridging and supplementary irrigation improves the moisture status of the soil and as a result buffers against extremes in moisture and reduces stress (heat and drought), which resulted in normal flowering, pod development as well as kernel developments. In line with this it had been reported that Pod and kernel development are progressively inhabited by drought stress due to insufficient plant turgor and lack of assimilates. Pod and kernel development may also be delayed by lack of soil water in pod zone (Boote and ketring, 1991; Strling and Black, 1991). It was also reported that number of pods per plant can be low due to increasing soil resistance (dryness of the soil as a result of moisture stress which resulted in difficulty of pegs to penetrate in to soil for pod formation) caused by prolonged drought (Sharma and Sivakumar 1991).

4.3.2.2 Number of seeds per pod

Analysis of variance indicated that the different integrated soil moisture conservation and soil fertility amendment management options did not affect number of seeds per pod at Hadinet and Lemlem sites (Table 4.5). The main reason for the non significance among the different management options might be due to the genetic potential of the variety used during the study. It has been reported that number of seeds per pod is under a genetic control (Ashley, 1984), even though environment and internal competitions might have an influence.

Table 4.5   Effect of integrated agronomic practices on yield attributes of groundnut

Treatments	Hadinet			Lemlem		
	Pods/plant (No.)	Seeds/pod (No.)	100 Seed wt(gm)	Pods/plant (No.)	Seeds/pod (No.)	100 Seed wt(gm)
C	5.97a	2.2	34.67a	8.33a	2.2	34.07a
F	15.00 bcd	2.7	46.50bc	18.47bcde	2.5	39.30b
GP	14.73 bcd	2.7	46.10b	17.93 bcde	2.7	41.00bcd
GP + F	12.07b	2.4	46.47bc	16.53b	2.8	**42.30 bcdef**
GP + SI +TD	20.00efg	2.6	51.93d	20.40 def	2.6	43.30 def
GP +SI	18.40def	2.5	50.13cd	19.60 cdef	2.8	44.43ef
GP +SI + F	16.27 bcdef	2.3	48.30bcd	17.40bcd	2.4	42.20 bcdef
GP +TD	11.93b	2.5	46.80bc	16.00b	2.7	43.83 def
GP +TD + F	13.33bc	2.7	45.80b	16.53b	2.7	41.57bcde
SI	13.87bc	2.8	50.93d	16.87bc	2.4	44.20ef
SI + F	16.07bcde	2.7	46.50bc	16.00b	2.6	40.03bc
SI + TD	17.57 cdef	2.4	49.10bcd	20.67ef	2.5	45.07f
SI +TD + F	23.53g	2.5	51.13d	16.00b	2.5	41.00bcd
SI +TD + F + GP	20.53fg	2.4	49.73bcd	21.80f	2.7	43.00 cdef
TD	13.27bc	2.4	48.47 bcd	18.33bcde	2.7	44.80f
TD + F	13.07b	2.5	48.70 bcd	19.87cdef	2.5	43.73def
DMRT[a] 0.5	4.377	ns	4.033	3.066	ns	3.196
CV %	17.1	8.2	5.1	10.5	8.2	4.6

C=Control, F=Fertilizer (DAP), GP=Gypsums, SI=Supplementary Irrigation, TD =Tied ridge
[a] Any means having the common letter are not significantly different at the 5% level of significance

4.3.2.3 Hundred Seeds weight

Significant difference (p < 0.001) was observed among treatments in terms of hundred seed weight at both locations (Table 4.5). The highest 100 seed weight (51.93gm and 45.07 gm) was recorded in gypsum + supplementary irrigation + tied ridge treatment combinations at Hadinet and supplementary irrigation + tied ridge treatment combinations at Lemlem sites, respectively (Table 4.5). In both sites the lowest 100 seed weight was observed in control. In both locations the integration of both moisture conservation practices with soil fertility amendment practices showed a better improvement on 100 seed weight of groundnut than the single application of management practices. But no superior advantage was observed between the moisture conservation practices (tied ridging and supplementary application) at both locations (Table 4.5). These values of 100 seed weight agreed with findings of Mukhtar (2011), who reported that the values of 100 seed weight of groundnut ranged between 45.92 to 50.41 gm in which different groundnut varieties were evaluated under different population density and basin sizes.

4.3.3 Dry pod yield and kernel yield

Analysis of variance indicated that a significantly different dry- pod and kernel yield ($p < 0.01$ at Hadinet; P=0.03 at Lemlem)   was recorded among treatments (Table 4.6).

The highest dry- pod yield and seed yield of (1941 and 980 Kg/ha, respectively) were found in treatment combination of supplementary irrigation + tied ridge + fertilizer (DAP) at Hadinet, while the highest pod and seed yield (2037 and 941.kg/ha, respectively) was found at Lemlem in the treatments supplementary irrigation and application of gypsum + supplementary irrigation, respectively (Table 4.6). No significant difference in yield was observed between the control and the DAP treated plots at Lemlem. While at Hadinet application of DAP showed significant effect over the control (Table4.6). Application of gypsum as source of calcium showed a significant effect in yield of pod in both locations alone and in combination with all management options.

Gypsum as a source of calcium is well documented and is widely used as a source of Ca for groundnut worldwide though it depends on the fertility status of the soil (Mupangwa. and Tagwira, 2005). The dissolution of gypsum is fairly rapid and therefore readily adds Ca to the podding zone. However, the major disadvantage of gypsum is its vulnerability to leaching especially on light textured soils.

The result showed that the combined effect of gypsum, fertilizer with tied ridging and supplementary irrigation practices are best options for enhancing production and productivities of groundnut at both experimental locations. The effect of soil fertility and moisture retention enhancement on groundnut is well studied. Accordingly, the result of this finding agrees with the work of Yebio et al. (1987) who reported that pod yield under irrigation ranged from 3.5 tons to 6.5 ton /ha. According to De et al (2005), ridge planting method not only maintained slightly higher soil moisture (8.4%) compared to the flat planting method (7.3%) but also produced higher kernel yield (0.57 t ha-^{1}) than flat planting (0.42 t ha-^{1}). In similar manner a number of studies had been conducted on the role of tied ridge in moisture conservation thereby enhancing yield of crops. Tied ridges have been found efficient in storing the rain water, which has resulted in substantial grain yield increase in some of the major dryland crops like sorghum, maize, wheat and Mung beans in Ethiopia. The average grain yield increased from 50 to 100 percent in tied ridge and traditional practice, respectively (Georgis and Takelle, 2000), though this increase will vary according to soil type, slop, rainfall and the crop grown.

From the result it can be suggested that the application of supplementary irrigation combined with soil fertility amendment practices at later growing season of groundnut maximizes water extraction and relatively better reducing of soil evaporation in comparing tied ridging. Thus both tied ridging and supplementary irrigation practices could buffer stress at late season, rainfall variability and distribution improving soil water there by affecting yield response.

Table 4.6 Effect of agronomic management options on dry pod and seed yield of groundnut

Treatments	Location					
	Hadinet			Lemlem		
	Dry Pod wt/ha(Kg)	Seed Yield/ha(kg)	Shelling (%)	Dry Pod wt/ha(Kg)	Seed Yield/ha(kg) (kg)	Shelling (%)
C	772a	290.a	0.378	1142.a	485a	0.4268
F	1481cdeb	543abc	0.374	1420.abc	642abc	0.4553
GP	1327bcd	671bcdef.	0.478	1759.bcde	845.bcd	0.4799
GP + F	957ab	612abcd	**0.768**	1420abc	641.abc	0.4460
GP + SI +TD	1543cdef	950.ef	0.628	1728.bcde	876.cd	0.5030
GP +SI	1728def	911.def	0.491	1790.cde	959.d	0.5371
GP +SI + F	1512cde	694bcdef	0.459	1512.abcd	755.abcd	0.4996
GP +TD	1296bc	683.bcdef	**0.517**	1759.bcde	880.cd	0.4877
GP +TD + F	1420cde	536. ab	0.376	1327.ab	585.ab	0.4375
SI	1358bcde	748.bcdef	0.569	2037e	941.d	0.4651
SI + F	1636cdef	751.bcdef	0.450	1574.abcd	708.abcd	0.4520
SI + TD	1728def	659bcdef	0.381	1759.de	848d	0.4752
SI +TD + F	1941f	980.f	0.490	1423.abc	640.abc	0.4445
SI +TD + F + GP	1543ef	863.cdef	0.486	1698.bcde	854.bcd	0.5017
TD	1420cde	654bcde	0.452	1790cde	889.cd	0.4924
TD + F	1265bc	601abcd	0.476	1605bcde	824bcd	0.5128
DMRT[a] 0.5	402.8	325.2	Ns	441.6	284.6	Ns
CV %	16.7	28	38.9	16.4	21.9	12.4

C=Control, F=Fertilizer (P), GP=Gypsum, SI=Supplementary Irrigation, TD =Tied ridge
[a] Any means having the common letter are not significantly different at the 5% level of significance

4.3.4 Dry Biomass yield

Statistical analysis of data revealed that the application of DAP, gypsum and supplementary irrigation alone and in double combination of each other showed a significant effect (P < 5%) over the control (Table 4.7). Across all the combination treatments no consistence was observed. Application of GP + F, GP + TD + F and TD practices alone did not show a significant effect over the control (Table 4.7). The highest and lowest dry biomass yield (4129 and 2043 kg/ha) was recorded from plots that received supplementary + tied ride + fertilizer (DAP) + gypsum and the control at Hadinet, respectively. At Lemelem, the highest dry biomass yield (3664kg/ha) was recorded in management options where gypsum + supplementary irrigation were implemented. Mean while, both the soil fertility amendments (DAP and Gypsum); and moisture conservation practices (SI and TD) did not statistically differ in dry biomass yield when tested without integration with each other (Table 4.7).

Table 4.7 Effect of integrated Agronomic management options on dry Biomas weight and Harvest index of groundnut

Treatments	Location				
	Hadinet			Lemlem	
	Dry biomass	HI (%)		Dry biomass weight/ha	HI (%)
C	2043.a	0.145		2348a	0.2058
F	3136.bc	0.182		3451def	0.1867
GP	3068.bc	0.207		3177bcde	0.2665
GP + F	2272ab	0.317		3300cdef	0.1931
GP + SI +TD	3718.cd	0.253		3277cdef	0.2662
GP +SI	3734.cd	0.229		3664f	0.2615
GP +SI + F	3264.cd	0.210		2941bc	0.2568
GP +TD	3098.bc	0.218		2790b	0.3295
GP +TD + F	2881.abc	0.184		2842b	0.2063
SI	3216bcd	0.230		3085bcd	0.3096
SI + F	3389.cd	0.218		2853b	0.2469
SI + TD	4092.d	0.165		3161bcde	0.2885
SI +TD + F	4118.d	0.238		2327a	0.2717
SI +TD + F + GP	4129.d	0.212		3540ef	0.2403
TD	2973.abc	0.216		3138bcd	0.2827
TD + F	3116bc	0.191		3303cdef	0.2504
DMRT[a] 0.5	944.7	ns		388.3	Ns
CV %	17.3	35		7.6	24.8

C=Control,   F=Fertilizer   (P),   GP=Gypsums,   SI=Supplementary   Irrigation,   TD   =Tied ridge,HI=harvest index
[a] Any means having the common letter are not significantly different at the 5% level of significance

4.3.5 Harvest index

Harvest index explains the relationship between grain yield (economic yield) and total biological yield and reflects the state of dry matter partitioning into grain yields. There was no significant difference in HI among the treatments at both sites (Table 4.7).

As indicated in Table 4.5 number of seeds/pod and number of pod/ plant are very low, thus this can result in marked lower harvest index.

4.4 Identification and incidence of Aspergillus species

Seeds collected from the experimental areas produced through application of integrated management options were examined for identification and incidence of fungus. Aspergillus were isolated from groundnut seed samples from both experimental sites. These include *A. flavus* and *A. niger* in both experimental sites. The incidence of *A. flavus* was 93 % and 100 % at Hadinet and Lemlem, respectively (Appendix Table 28). The incidence of *A. niger* was 100 % and 93 % at Hadinet and Lemlem, respectively (Appendix 30). Abera (2009 unpablished data) also reported that species of Aspergillus were dominant in the area. *Aspergillus flavus,* however was, the dominant fungi (Table4.8). This result was also in agreement with the findings of Indress et al (2010); Sultan and Magan (2010) which showed that *A. flavus* was the main species occurring in peanut seed.

4.5 Effect of integrated agronomic practices on infection of groundnut seed by Aspergillus species

Groundnut seed infection by Aspergillus species can occur through various conditions. Aspergillus infection can occur under pre and post harvest handling and storage conditions (Mehan et al., 1991). A number of management options had been practiced for minimizing aflatoxin contamination caused by Aspergillus species. Breeding of resistant varieties, good agricultural production, processing, and handling and storage practices are some of the different management options. But it has been reported by Waliyar et al. (1994) that little success was achieved in the development of groundnut varieties with resistance to aflatoxin contamination. In line with this an experiment was conducted using different management options. Results of the different integrated agronomic management options in minimization of Asperggillus species infection on groundnut are presented below.

4.5.1 Effect of soil fertility amendment practices on Asspergillus species infection

Application of fertilizer (DAP) at planting as a means of reducing *A.flavus* and *A.niger* infection of groundnut showed a significant effect   (P< 5%) over the control at Lemlem   (Table 4.8). Significance difference on application of gypsum as a source of calcium over control was observed at both experimental site. From this result it can be suggested that the role of Calcium in reducing aspergillus infection is highly emphasized. Applications of calcium in groundnut production alleviate problems like aborted and shriveled kernel seed of peanut and high percentage of 'pops' which are major sources for aspergillus infection. Florence (2011) reported that application of gypsum as a source of calcium is crucial to have good quality with totally sound matured groundnut seed

4.5.2 Effect of Moisture retention Amendment Practices and Asspergillus species Infection

Analysis of variance indicated that the integrated management options of soil moisture conservation practices applied at the early flowering stage resulted in a significant effect (P< 5%) in *Aspergillus flavus* infection at both experimental areas. However, the integrated management options did not affect the level of infection in *A. niger* at both locations. The lowest infection levels (3.0 and 4.3%) were recorded in SI + TD at Hadinet and Lemlem, respectively. But the highest severity levels (17.3 and 19.3%) were recorded in control at Hadinet and Lemem experimental site, respectively (Table 4.8).

Table 4.8. Effect of integrated Agronomic practices on A.flavus, and A.nige infection (severity) of groundnut seed

Treatments	Hadinet		Lemlem	
	A.flavus Infection (%)	A.niger Infection (%)	A.flavus Infection (%)	A.niger nfection (%)
C	17.3d	5.7	19.3f	6.3
F	9.7bcd	6.3	7.7e	5.0
GP	7.0abc	5.0	8.0e	4
GP + F	10.0cd	5.3	7.3de	2.0
GP + SI +TD	4.3abc	3.0	4.7ab	4.0
GP +SI	6.7abc	3.3	5.7abc	2.3
GP +SI + F	10.3cd	4.0	5.0ab	1.7
GP +TD	5.7abc	3.7	7.3de	2.3
GP +TD + F	10.0bcd	3.3	8.0e	3.7
SI	4.0ab	2.7	6.0bcd	2.3
SI + F	5.0abc	2.3	6.7cde	2.7
SI + TD	3.0a	1.3	4.3a	2.7
SI +TD + F	6. 0abc	3.7	5.0ab	1.7
SI +TD + F + GP	10.3cd	6.3	5.3abc	2.3
TD	7.7abc	2.7	7.7e	4.0
TD + F	7.3abc	6.0	5.0ab	2.7
DMRT[a] 0.05	1.238	ns	0.2918	ns
CV %	27.5	33.5	6.5	32.5

[a] Any means having the common letter are not significantly different at the 5% level of significance

This result also agreed with findings of Cole et al (1989) who found that there was higher infection of peanut by Aspergillus and subsequent higher contamination where there was moisture stress during the last 20 -40 days to growth period of the crop. In similar manner the role of irrigation on minimizing Aspergillus infection were highly emphasized. Finding on irrigation frequency and sowing date on groundnut reveals that infection of Aspergillus increased as the frequency declined from 7 day to non (rainfed) (Craufard et al.,2004). As indicated in (Table 4.8) no significance difference in *A.flavus* infection was observed between the two soil moisture amendments options (tied and supplementary irrigation) managements options implemented. In addition combination of all the soil moisture retention amendment practices (TD and SI) had no any significant difference as compared with the double and triple combination of soil fertility amendment practices (Table 4.8).Un like Lemlem application of fertilizer, gypsum + fertilizer, gypsum + supplementary + fertilizer, gypsum + tied + fertilizer and   supplementary irrigation + fertilizer , gypsum + tied ridge combinations did not showed a significant over the control (Table 4.8). This might occur due to termite a problem which is serious problem in the area other factors which result in moisture difference in the experimental plots.

4.6 Moisture content at harvesting and Aspergillus infection of groundnut

Statistical analysis indicated that moisture content and Aspergillus infection have inverse relations (Figure 4.1 and 4.2). Moisture content increased from 3% at control to 9.04 % in supplementary irrigation + tied ridge + gypsum combinations at Hadinet (Figure4.1). The infestation level of Aspergillus decreased from 17% at control to 3% in supplementary irrigation and tied ridge combination at Hadinet (Figure 4.1). In Lemlem the moisture content ranged from 4.32 % in control to 14.63 % in supplementary irrigation, tied ridge, Fertilizer (DAP) and gypsum combinations. As a result the infection level of Aspergillus decreased from 19.3% in control (without intervention of any agronomic practices) to 4.3% in supplementary irrigation and tied ridge combination (Figure 4.2).

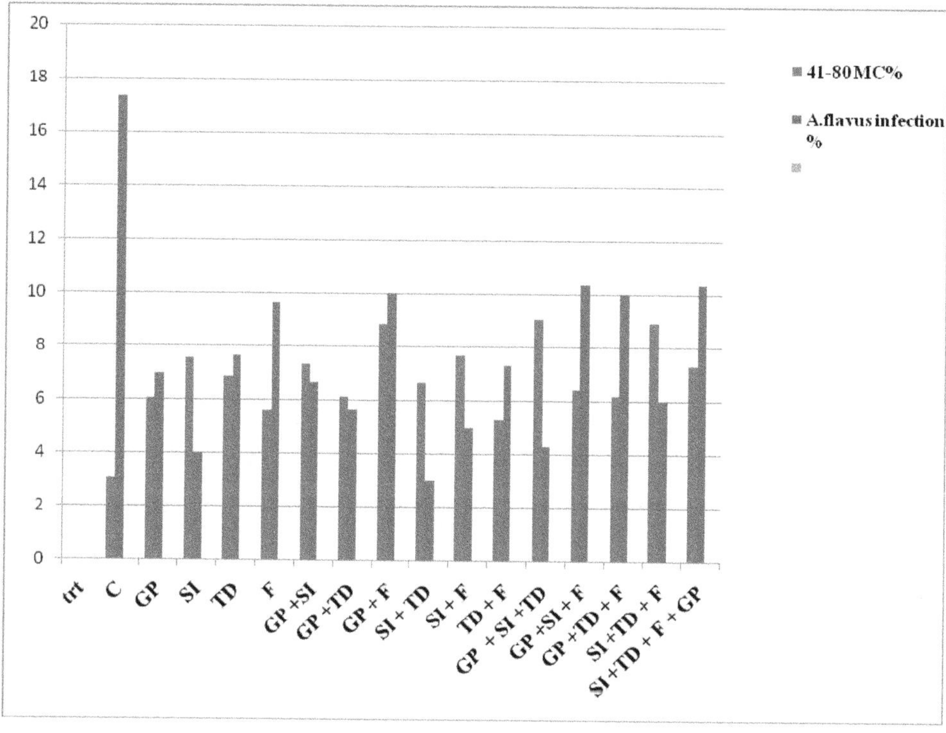

Figure 4.1 Moisture content retention and Aspergillus infecton in groundnut at 41-80 cm root depth at Hadinet

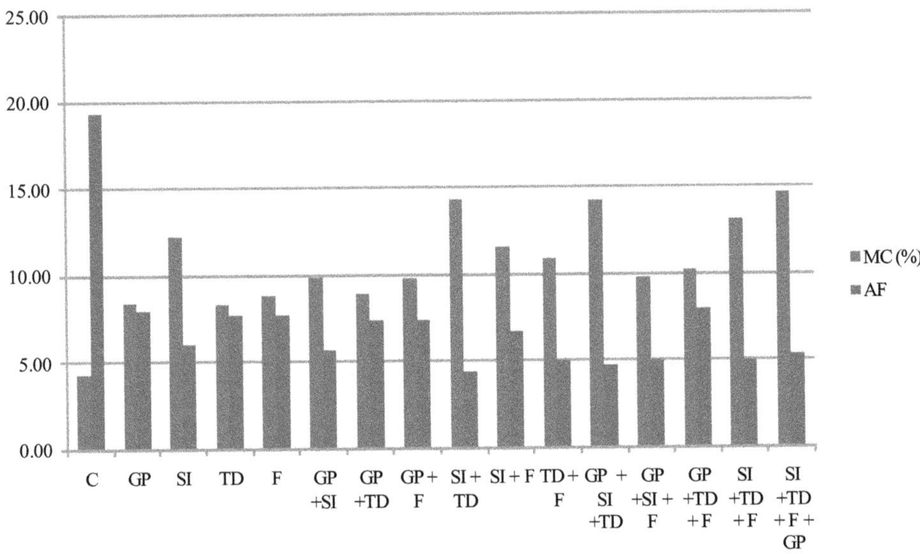

Figure 4.2 Moisture content retention and Aspergillus infection in groundnut at 41-80 cm root depth Lemlem

Chapter five: Conclusion and Recommendation

Conclusion

From the soil fertility amendment management practices gypsum application considerably increased dry pod and kernel yield of groundnut by 72 – and 131 % over the control at Hadinet,, respectively. While 54 and 74% of dry pod and kernel yield increment over the control was recorded at Lemlem experimental site, respectively.

Application of gypsum as a source of calcium at early flowering stage significantly lowered Aspergillus infection compared over the control at Hadinet and lemlem experimental site

Application of DAP fertilizer significantly lowered Aspergillus infection only at lemlem experimental site indicating that moisture has a direct relationship with Aspergillus infection .

Application of soil moisture conservation practice (tied ridging) at early flowering stage of groundnut and supplementary irrigation at the cessation of rain fall significantly lowered Aspergillus infection compared to control at Hadinet and lemlem experimental site. At Hadinet experimental site no significant effect was observed between tied ridge and supplementary irrigation application as means of reducing aspergillus infection indicating that based on the availability one of them independently can serve as a means of reducing aspergillus infection.

Though there was no consistent in reducing the infestation level of *Aspergillus flavus* over all agronomic combination practices, combination of supplementary irrigation, tied ridge, fertilizer and gypsum reduced the level of *Aspergillus flavus* infestation by 68% over the control.

Application of supplementary irrigation together with tied ridging and fertilizer significantly improved dry pod yield and kernel yield of groundnut at Hadinet experimental site. Supplementary irrigation and tied ridge combination was effective and consistent (in comparing the other agronomic practices) agronomic practices that have significantly reduced level of Aspergillus infection over the control.

Seeds samples were infested by *Aspergillus flavus* and *Aspergillus niger*. *Aspergillus flavus* was the dominantly occurred fungus in both locations .

Recommendation

Soil moisture and fertility amendment practices reduce the level *Aspergillus flavus* infection level but to what level of aflatoxin these agronomic practices did reduce has to be considered for future research direction in verifying the present investigation across years and locations.

The experiment showed that in areas where water for supplementary irrigation is not available application of tied ridge becomes best alternative option.

6. References

Abdalla, A.T., C.J. Stigter, H.A., Mohamed, A.E., Mohammed and M.C. Gough. (2005). Identification of micro-organisms and mycotoxin contamination in underground pit stored sorghum in central Sudan.

Abera (2009).Infection severity of Aspergillus spp and aflatoxin contamination in Tigray region. Un published data.

Ae, N., Otani, T., Makino, T. and Tazawa, J. (1996). Role of cell wall of groundnut roots in solubilizing sparingly soluble P in soil. Plant Soil 186: 197–204.

Alireza, H. G., Mohammad, N. S. V. and Mohammad, H. H. (2012). Effect of Potassium and Calcium Application on Yield, Yield Components and Qualitative Characteristics of Peanut (*Arachis hypogaea* L.).World Applied Sciences Journal 16 (4): 540-546.

Allen, R.G., Pereira, L.S., Raes, D. and Smith, M. (1998). Crop evapotranspiration: guideline for computing crop water requirement. Irrigation and drainage paper number 56.FAO, Rome.

Ashley, J.M. (1984). Groundnut, Pp 467. In: Peter, R.G.and Fisher, N.M. (eds.). The Physiology of Tropical Field crops. A Wiley- Interscience Publication.

Assmo, P. and Eriksson, A. (1999). Soil conservation in Arusha Region, Tanzania. Manual for Extension.Technical Handbook No. 7. RELMA, Nairobi.

Bennett, J.w. (2010). An Overview of the Genus Aspergillus. In: Aspergillus: Molecular Biology and Genomics, Machida M. and Gomi K. eds.

Bhatia, V.S., Singh, P., Wani, S.P., Kesava, R. A.V.R. and Srinivas, K. (2006). Yield Gap Analysis of Soybean, Groundnut, Pigeonpea and Chickpea in India Using Simulation Modeling. Global Theme on Agroecosystems Report no. 31. Patancheru 502 324, Andhra Pradesh, India: ICRISAT, 156 pp.

Boote, K.J. and Kettering, D.L. (1991). Peanut . IN: Stewurt B.A. and Nielson D.R.(eds). Irrigation of Agricultural crops . ASACSSA- SSSA. Madson.

Brennman,T.B.et al.(1995). Suppression of foliar and soilborn peanut diseases in bahiagrass rotation . phytopathology,85(9), 948-52. In: Weiss, E. A.(2000). Oil seed crops. World Agriculture Seris, 2nd edition, 65p

Brereton, R.G. (1980). Groundnut: A case with great potential for Ethiopia. Ethiopian Grain Review, 6: 22-23.

Chikowo, R. (1998). Soil fertility management options for improved groundnut production in the smallholder sector.M.Phil. Thesis, University of Zimbabwe.In: Mupangwa, W.T and. Tagwira, F(2005). Groundnut yield response to single uperphosphate, calcitic lime and gypsum on acid granitic sandy soil. Nutrient Cycling in Agro ecosystems. 73:161–169.

Cole, R.J. and Cox, R.H. (1981). Handbook of Toxic Fungal Metabolites (New York: Academic Press). In: Bennett J.w, (2010). An Overview of the Genus Aspergillus.

Cole, R.J., Sanders, T.J., Dorner, J.W., Blankenship, P.D. (1989). Environmental conditions required to induce pre-harvest concentration in groundnut. Summary of six years research.

Craufurd, P.W., Prasad, P.V.V., Waliyar, F. and Taheri, A. (2006). Drought, pod yield, pre-harvest Aspergillus infection and aflatoxin contamination on peanut in Niger. Field Crops Research,98: 20-29.

Critchley, W., Siegert, K. (1991).Water Harvesting. FAO, Rome. IN: Mati, B. M. 2005. Overview of water and soil nutrient management under smallholder rainfed agriculture in East Africa. Working Paper 105 Colombo, Sri Lanka: International Water Management Institute (IWMI).

CSA. (2010). Area and production of Crops. Vol, I statistical bulletin No. 446, Addis Abeba, Ethiopia.

Das, S.K., Rao, A.C.S. and Sharma, K.L.(1991).Legume based crop rotations on dryland Alfisol. Indian Journal of Dryland Agriculture, Research Dev. 6, 46±59.

David, L.J. (2011).Peanut production practices.IN: Davied et.al. (eds) (2011). Peanut information North Carolina Cooperative Extension Service.NC state University pp.36.

De ,V. J.W., Trucksess, M.W. and Jackson, L.S. (2002). Mycotoxins and Food Safety (New York: Kluwer Academic/Plenum Pubs.). **In**: Bennett J.w, (2010). An Overview of the Genus Aspergillus.

De, P., Chakravarti, A.K., Chakraborty, P.K. and Chakraborty, A. (2005). Study on the efficiencyof some bio-resources as mulch for soil moisture conservation and yield of groundnut (*Arachishypogaea* L.), Archives of Agronomy and Soil Science, 51: 247-252.

Desai, B.B., Kotecha, P.M. and Salunkhe, D.K. (1999). Science and Technology of Groundnut: Biology, Production, Processing and Utilization. Naya Prokash, Calcutta, 677 pp. Ethiopian Agricultural research Institute (2007). Manual for the improved technologies. Addis Ababa , Ethiopia.

Diener, U. L. and Cole, R. J. (1982). Aflatoxins and other mycotoxins in peanuts. In: Peanut science and Technology Pattee, H. E. & Young, C.T. Eds. Yoakum, Texas USA. American Peanuts Research and Education Society: 486-519.

Dorner, J .W. (2007). Management and prevention of mycotoxins in peanuts. Food Additives and Contaminants, 25 (2) : 203-208.

Edlayne, G., Juliana, H., Nogueira, H., Jonan, D., Francisco, A. and Benedito, C. (2008). Mycobiota and mycotoxins in Brazilizn peanut kernels from sowing to harvest. International Journal of Food Microbiology, 23:184 – 190.

Erick, L. (2009).Minimizing Alfatoxine in Corn. Mississippi state university Extension service.

FAO. **(2003)**. Statistical databases. http://www.FAO.ORG

Florence, R. J. (2011). Fertilization of Peanut (*Arachis hypogaea* L.) with Calcium: Influence of Source, Rate, and Leaching on Yield and Seed Quality. Msc. thesis, Auburn University, Auburn, Alabama.

Fana, S. (2010). Revitalization of the Groundnut Sector in West Africa (Gambia, Guinea Bissau and Senegal). Global Agricultural Network, USDA Foreign Agricultural Service.

Faujdar, S. and Oswalt, D.L. (1995). Groundnut Production Practices. ICRISAT, Skill evelopment Series no. 3 Revised.

Freeman, H.A., Nigam, S.N., Kelly, T.G., Ntare, B.R., Subrahmaniyam, P. and Boughton, D. (1999). The World groundnut economy: facts, trends, and outlook. ISRISAT, 52 pp

Georgis, K. and Takele, A. (2000).Conservation farming technologies for sustaining crop production in semi-arid areas of Ethiopia.

Getinet, A., Geremew, T., Kassahun, Z. and Bulcha, W.(1997).Low land oil crop: a three – decade research experience in Ethiopia . Research Report No. 31.Institute of Agriculture Research , Addis Ababa, Ethiopia.

Gomez, A.K. and Gomez, A.A. (1984). Statistical Procedures for Agricultural Research. Second edition. John Wiley and Sons Inter-Science Publications. New York.

Gong, Y.Y, Cardwell, K., Hounsa, A., Egal, S., Turner, P.C., Hall, A.J. and Wild, C.P. (2002).Cross-sectional study of dietary aflatoxin exposure and impaired growth in young children from Benin and Togo, West Africa. British Medical Journal, 325, 20-21.

González, H., Resnik, S., Boca, R.. and Marasas, W.(1995). Mycoflora of Argentinean corn harvested in the main production area in 1990. Mycopathologia, 130: 29-36.

Henry, S.H., Bosch, F.X., Troxell, T.C. and Bolger, P.M. (1999). Reducing liver cancer – global control of aflatoxin. Science 286, 2453–2454. In: Bennett J.w, (2010). An Overview of the Genus Aspergillus.

Hossain, M. A, Hamid, A. and Nasreen, S. (2007*)*. Effect of nitrogen and phosphorus fertilizer on N/P uptake and yield performance of groundnut (*arachis hypogaea L).* Journal of Agricultural Research, 45(2).

Hulugalle, N.R. (1990). Alleviation of soil constraints to crop growth in the upland Alfisols and associated soil groups of the West African Sudan Savannah by tied-ridges. Soil Tillage Research 18, 231±248.

ICRISAT (2005).Aspergillus flavus seed infection and aflatoxin estimation by ELISA and Aflatoxin management in groundnut.

Idress, H.A.,Laith, T.A.A., Muftah, A.N., Ibrahim, A.A.B., Maziah, Z., Saad, E.M.,and Rezaul, K.S.M.(2010). Screening of fungi associated with commercial grains and animal feeds. World Applied Science Journal, 9 (7) : 746- 756.

Jackson, M. L. (1958). Soil chemical analysis. Prentice-Halls Inc., Madison, Wisconsin, USA, pp.545-566.

Jackson, M. L. (1967). Soil Chemical Analysis Practice Hall of India. New Delhi

Kamara, E.G., Olympia, N.S. and Asibuo, J.Y. (2011). Effect of calcium and phosphorus fertilizer on the growth and yield of groundnut (*Arachis hypogaea L.*).International Research Journal of Agricultural Science and Soil Science, 1(8): 326-331.

Kidane, G. (1999). Agronomic Techniques for Higher and Sustainable Crop Production in the Dryland Areas of Ethiopia: Food Security Perspective. In Food SecurityThrough Sustainable Land Use. Policy on Institutional, Land Tenure, and Extension Issues in Ethiopia, ed. T. Assefa. Proceedings of the First National Workshop of NOVIB Partners Forum on Sustainable Land Use. Addis Ababa, NOVIB. 99-115.

Kochert, G., Stalker, H.T., Gimenes, M.,Galaro, L., Lopes, C.R. and Moore, K.(1996). RFLP and cytogenetic evidence on the origin and evalution of allotetraploid domesticated peanut *Arachis hypogaea* (Leguminosae). American Journal of Botany, 83:1282-1291.

Kumar, K. and Rao, K. V. P. (1991). Nitrogen and phosphorus levels in relation to dry matter production, uptake and their partitioning in soybean. Ann Agricultural Research, 12:270-272.

Lal, R.. (1995).Tillage systems in the tropics, management options and sustainability implications. FAO Soils bulletin No. 71, Rome, Italy.

Legesse, Y.(1999). Agroecology of zone of Tigray land use planning division of Tigray Buaraue of Agriculture and natural resource.

Lewis, L., Onsongo, M., Njapau, H., Schurz-Rogers, H., Luber, G., Kieszak, S., Nyamongo , J., Backer, L., Dahiye, A.M., Misore, A., DeCock, K., Rubin, C. and the Kenya Aflatoxicosis Investigation Group (2005). Aflatoxin contamination of commercial maize products during an outbreak of acute aflatoxicosis in Eastern and Central Kenya. Environmental Health Perspectives, 113:1763-1767.

Liming, C. and Warren, A.D. (2011).Gypsum as an Agricultural Amendment. General Use Guide lines. The Ohio State University.

Mati, B. M. (2005). Overview of water and soil nutrient management under smallholder rainfed agriculture in East Africa. Working Paper 105. Colombo, Sri Lanka: International Water Management Institute.

Mehan, V.K., Mayee, C.D., Jayanthi, S. and McDonald, D. (1991). Pre-harvest seed infection by Aspergillus flavus group of fungi and subsequent aflatoxin contamination in groundnuts in relation to soil types. Plant Soil, 136:239-248.

Michael, Z.Z., John, L.R. and Johann, B. (2006). A review of rapid methods for the analysis of mycotoxins. Mycopathologia,, 161: 261–273.

Mirvat, E. G., Magda, H. M. and Tawfik, M.M. (2006). Effect of Phosphorus Fertilizer and Foliar Spraying with Zinc on Growth,Yield and Quality of Groundnut under Reclaimed Sandy Soils. Journal of Applied Science Research, 2(8): 491-496.

Mukhtar, A.A. (2011).Intensifying groundnut production in Sudan savanna zone of Nigeria including in irrigated croppying system. Pakistane Journal of Biological science, 14 (22):1028 – 1031.

Mupangwa, W.T. and Tagwira, F. (2005). Groundnut yield response to single superphosphate calcitic lime and gypsum on acid granitic sandy soil.

Nguyen, C. V. (2003). Phosphorus Fertilization Strategies for Groundnut Grown on Upland Acid Soils in Nghe An Province Better Crops International, 17(2).

Nigam, S.N., Giri, D.Y. and Reddy, A.G.S.(2004). Groundnut seed production manual. Patancheru 502 324, Andhra Pradesh, India: International Crops Research Institute for the Semi-Arid Tropics, 32 pp.

Norman, M.J.T., Pearson, C.J. and Searle, P.G.E.(1996). Tropical food and their environment. Cambridge University Press, 201-205p.

Ntare, B.R., Waliyar, F., Ramouch, M., Masters, E. and Ndjeunga, J (eds.) (2005). Market prospects for Groundnut in West Africa (Eng/Fr). CFC Technical Paper No. 39 PO Box 74656, 1070 BR Amsterdam. The Netherlands: Common Fund for Commodities; and Patancheru, India: International Crops Research Institute for the Semi-Arid Tropics, 252 pp.

Okello, D., Kaaya, A., Bisikwa, J., Were, M. and Oloka, H. (2010). Management of Aflatoxins in Groundnuts: A manual for Farmers, Processors, Onwueme I.. and Sinha T.(1999).Field Crop Production in Tropical Africa.

Olsen, S. R. and L. A. Dean. (1965). Phosphorous .*In:* Methods of Soil Analysis. American society of Agronomy. Maidison, Wisconsion .9:920-926.

Onwueme, I.C. and Sinha, T.I. (1991). Field crop production in tropical Africa. The Netherlands, CTA.

Otsuki, T., Wilson, J.S. and Sewadeh, M. (2001). What price precaution? European harmonization of aflatoxin regulations and African groundnut exports. European Revie of Agricultural Economics, 28 (3): 263-83.

Pitt, J.I., Hocking, A.D. and Glenn, D.R. (1983). Identification and characterization of *Aspergillus flavus* and aflatoxins. The Journal of Applied Bacteriology, 54: 109 .

Rachaputi, N., Wright G.C., and. Krosch, S. (2002). Management practices to minimize pre-harvest aflatoxin contamination in Australian peanuts. *Aust.* J. Exp. Agric., 42(5):595–605.

Selvaraju, R. Subbian, P. Balasubramanian, A. and Lal, R. (1999).Land configuration and soil nutrient management options for sustainable crop production on Alfsols and a Vertisols of southern peninsular India. ELSEVIER, Soil & Tillage Research, 52: 203 – 216.

Sene, M. and Badiane, A.N.(2005). Nutrient and water management practices for increasing Crop production in rainfed Arid/Semi-Arid Areas, optimization of water and nutrient use by maize and peanut in rotation based on organic and rock phosphate soil amendments. International atomic energy agency. Proceedings of a coordinated research project.

Sharma, P.S. and Sivakumar, M.V.K. (1991). Penetrometer soil resistance: Pod number and yield of peanuts as influenced by drought stress. Indian Journal Plant Physiology, 34 (2): 147–152.

Smartt, J.J.,Gregory W.C. and Gregory M.P. (1978). The genomes of *Arachis hypogaea*, *Euphytica* ,27:665-675.

Smith, W.C. (1995). Crop Production: Evolution, History and Technology. John Wiley & Sons, INC, pp. 408–457.

Steven, J. and Luz, B. D. R.(2008). Agriculture and Rural Development. Barrier, Catalyst, or Distraction Standards, Competitiveness, and Africa's Groundnut.

Stirling, C.M. and Black, C.R. (1991). Stages of reproductive development in groundnut (*Archis hypogaea* L.) most susceptible to environmental stress. Tropical Agriculture (Trinidad), 68(3): 296–300.

Sultan, Y.and Magan, N. (2010).Mycotoxigenic fungi in peanut from different geographic region of Egypt.Mycotoxin Resarch ., 26: 133 -140.

UNIDO (2010).Capacity Building For Aflatoxin Management In Groundnuts In Malawi . Ministry Of Industry and Trade, And Ministry Of Agriculture and Food Security: Government Coordinating Agency, Final Report February 2009 - December 2010.

Waliyar, F., Ba ,A., Hamma, H., Bonkoungou, S. and Bosc, J.P. (1994). Sources of resistance to Aspergillus flavus and aflatoxin contamination in groundnut genotypes in West Africa. Plant Disease, 78: 704-708.

Waliyar, F., Ntare, B.R., Diallo, A.T., Kodio, O. and Diarra, B. (2007). On-farm Management of Aflatoxin Contamination of Groundnut in West Africa. A Synthesis Report. ICRISAT, 24 pp.

Weidenborner, M. (2001) .Encyclopedia of Food Mycotoxins (Berlin: Springer).

Weiss, E. A. (2000). Oil seed crops. World Agriculture Seris, 2nd edition, 65p.

Weiss, E. A. (2000). Phosphatic fertilizer usage on West Kenya crops. Agricultural Digest, 18, 29-38.In: Weiss, E. A.(2000). Oil seed crops. World Agriculture Seris, 2nd edition, 65p.

Williams, J.H., Phillips, T.D., Jolly, P.E., Stiles, J.K., Jolly, C.M. and Aggarwal, D. (2004). Human aflatoxicosis in developing countries: a review of toxicology, exposure, potential health consequences, and interventions. Am. J. Clin. Nutr. 80, 1106–1122.

Wright, G. (2004). Peanut. IN: Colin,w.,Horald,C. and Charles,E.W(eds). The Encyclopedia of Grain Science, 438-445. ELSVER academic press.

Yebio, W., Seme, D., Asfaw, Z., Amare, A., Abebe, T. and Beniwal, S. P. (1987). Research on groundnut, pigeon pea and chickpea in Ethiopia.

Zharare, G.E. (1996). Research priorities for groundnut (Arachis hypogaea L.) nutrition- A scientific basis for manipulating soil fertility to optimize groundnut yields. Agronomy Institute Annual Review and Planning Workshop, 6–7 August 1996. Department of Research and Specialist Services,Harare, Zimbabwe.

Zharare, G.E., Asher, C.J., Blamey, F.P.C. and Dart, P.J. (1993). Pod development of groundnut A. hypogaea L.) in solution culture. Plant Soil 155/156: 355–358. **In**: Mupangwa, W.T and. Tagwira, F(2005). Groundnut yield response to single uperphosphate, calcitic lime and gypsum on acid granitic sandy soil. Nutrient Cycling in Agro ecosystems, 73:161–169.

Appendix

Apendix Table 1. Crop evapotrspiraton and crop water requirement

methrological data	country Ethiopia		13 14 06 N
		Latitude ,13.25 N	
		Longitude ,38.98 E	38 58 50 E
Altitude 1500m	station , Abergele		

month	Mini Temp °C	Max Temp °C	Humidity %	wind m per sec	SS hurs	RF mm	Pitch Evaporimeter mm
Jan	13.13	31.47	37.67	2.39	9.1	1.32	11.15
Feb	14.87	33.73	31	2.7	8.9	1.08	12.25
March	15.03	34.6	31.67	2.1	8.5	15.44	12.3
April	16.17	35.57	30.67	2.74	7.6	32.88	15.55
May	16.72	36.13	31.33	2.59	7.7	16.22	15.15
June	14.73	33.4	36	2.73	6.6	74.7	16.5
July	13.23	31	40.33	2.3	4.2	118.32	11.5
August	12.4	30.13	49.67	1.87	3.5	186.03	5.2
Sept	13.13	31.6	41	1.72	4.4	68.13	9.7
Oct	14.6	32.9	34.67	2.16	8.2	5.93	12.5
Nov	13.95	31.95	35.67	2.37	10.2	10.67	11.8
Dec	13.48	31.2	36	2.37	9	0	ND

Appendix Table 2. Analysis of variance of 50% days to floworing at Hadinet experimental site

Source of variation	d.f.	s.s.	m.s.	v.r.	F pr.
Rep stratum	2	75.125	37.562	15.46	
Rep.*Units* stratum					
Mgt	15	69.813	4.654	1.92	0.063
Residual	30	72.875	2.429		
Total	47	217.812			

Appendix Table 3. Analysis of variance of maturity days at Hadinet experimental site

Source of variation	d.f.	s.s.	m.s.	v.r.	F pr.
Rep stratum	2	58.042	29.021	3.78	
Rep.*Units* stratum					
Mgt	15	156.812	10.454	1.36	0.230
Residual	30	230.625	7.688		
Total	47	445.479			

Appendix Table 4. Analysis of variance of plant height (cm) at Hadinet experimental site

Source of variation	d.f.	s.s.	m.s.	v.r.	F pr.
Rep stratum	2	58.796	29.398	3.83	
Rep.*Units* stratum					
mgt	15	694.383	46.292	6.03	<.001
Residual	30	230.430	7.681		
Total	47	983.610			

Appendix Table 5. Analysis of variance of dry biomass weight (kg/ha) at Hadinet experimental site

Source of variation	d.f.	s.s.	m.s.	v.r.	F pr.
Rep stratum	2	3579874.	1789937.	5.58	
Rep.*Units* stratum					
mgt	15	16258015.	1083868.	3.38	0.002
Residual	30	9627806.	320927.		
Total	47	29465695.			

Appendix Table 6. Analysis of variance of dry pod weight (kg/ha) at Hadinet experimental site

Source of variation	d.f.	s.s.	m.s.	v.r.	F pr.
Rep stratum	2	1459301.	729651.	12.50	
Rep.*Units* stratum					
mgt	15	3979579.	265305.	4.55	<.001
Residual	30	1750632.	58354.		
Total	47	7189513.			

Appendix Table 7. Analysis of variance of Harvest index (%) at Hadinet experimental site

Source of variation	d.f.	s.s.	m.s.	v.r.	F pr.
Rep stratum	2	0.082767	0.041384	7.28	
Rep.*Units* stratum					
Mgt	15	0.068110	0.004541	0.80	0.670
Residual	30	0.170639	0.005688		

| Total | 47 | 0.321517 | | | |

Appendix Table 8. Analysis of variance of Kernel yield (kg/ha) at Hadinet experimental site

Source of variation	d.f.	s.s.	m.s.	v.r.	F pr.
Rep stratum	2	1729367.	864684.	22.74	
Rep.*Units* stratum					
mgt	15	1376426.	91762.	2.41	0.019
Residual	30	1140937.	38031.		
Total	47	4246730			

Appendix Table 9. Analysis of variance of available soil moisture content (%) at 0- 40cm root depth at Hadinet experimental site

Source of variation	d.f.	s.s.	m.s.	v.r.	F pr.
Rep stratum	2	25.650	12.825	5.53	
Rep.*Units* stratum					
mgt	15	57.765	3.851	1.66	0.116
Residual	30	69.568	2.319		
Total	47	152.984			

Appendix Table 10. Analysis of variance of pods per plant at Hadinet experimental site

Source of variation	d.f.	s.s.	m.s.	v.r.	F pr.
Rep stratum	2	85.715	42.858	6.22	
Rep.*Units* stratum					
mgt	15	773.553	51.570	7.49	<.001
Residual	30	206.672	6.889		
Total	47	1065.940			

Appendix Table 11. Analysis of variance of seeds per pod at Hadinet experimental site

Source of variation	d.f.	s.s.	m.s.	v.r.	F pr.
Rep stratum	2	0.30167	0.15083	3.43	
Rep.*Units* stratum					
mgt	15	1.27667	0.08511	1.94	0.060
Residual	30	1.31833	0.04394		
Total	47	2.89667			

Appendix Table 12. Analysis of variance of   shelling percentage at Hadinet experimental site

Source of variation	d.f.	s.s.	m.s.	v.r.	F pr.
Rep stratum	2	0.38353	0.19176	5.38	
Rep.*Units* stratum					
mgt	15	0.47463	0.03164	0.89	0.583
Residual	30	1.06859	0.03562		
Total	47	1.92675			

Appendix Table13. Analysis of variance of *Aspergillus flavus* (transformed data) at Hadinet experimental site

Source of variation	d.f.	s.s.	m.s.	v.r.	F pr.
Rep stratum	2	15.2144	7.6072	13.80	
Rep.*Units* stratum					
mgt	15	17.4528	1.1635	2.11	0.040
Residual	30	16.5369	0.5512		
Total	47	49.2041			

Appendix Table 14. Analysis of variance of 50% days to flowering at Lemlem experimental site

Source of variation	d.f.	s.s.	m.s.	v.r.	F pr.
Rep stratum	2	168.042	84.021	23.35	
Rep.*Units* stratum					
mgt	15	61.917	4.128	1.15	0.361
Residual	30	107.958	3.599		
Total	47	337.917			

Appendix Table 15. Analysis of variance of maturity days at Lemlem experimental site

Source of variation	d.f.	s.s.	m.s.	v.r.	F pr.
Rep stratum	2	408.50	204.25	18.37	
Rep.*Units* stratum					
mgt	15	373.25	24.88	2.24	0.029
Residual	30	333.50	11.12		
Total	47	1115.25			

Appendix Table 16. Analysis of variance of plant height (cm) at Lemlem experimental site

Source of variation	d.f.	s.s.	m.s.	v.r.	F pr.
Rep stratum	2	163.527	81.763	17.32	
Rep.*Units* stratum					
mgt	15	117.490	7.833	1.66	0.116
Residual	30	141.620	4.721		
Total	47	422.637			

Appendix Table 17. Analysis of variance of dry biomass weight (kg/ha) at Lemlem experimental site

Source of variation	d.f.	s.s.	m.s.	v.r.	F pr.
Rep stratum	2	369481.	184740.	3.41	
Rep.*Units* stratum					
mgt	15	6480450.	432030.	7.97	<.001
Residual	30	1626611.	54220.		
Total	47	8476542.			

Appendix Table 18. Analysis of variance of dry pod weight (kg/ha) at Lemlem experimental site

Source of variation	d.f.	s.s.	m.s.	v.r.	F pr.
Rep stratum	2	1254215.	627108.	8.94	
Rep.*Units* stratum					
mgt	15	2460691.	164046.	2.34	0.023
Residual	30	2104049.	70135.		
Total	47	5818955			

Appendix Table 19. Analysis of variance of pods per plant (kg/ha) at Lemlem experimental site

Source of variation	d.f.	s.s.	m.s.	v.r.	F pr.
Rep stratum	2	2.647	1.323	0.39	
Rep.*Units* stratum					
mgt	15	425.339	28.356	8.39	<.001
Residual	30	101.433	3.381		
Total	47	529.419			

Appendix Table 20. Analysis of variance of seeds per pods (kg/ha) at Lemlem experimental site

Source of variation	d.f.	s.s.	m.s.	v.r.	F pr.
Rep stratum	2	0.00792	0.00396	0.09	
Rep.*Units* stratum					
mgt	15	1.19146	0.07943	1.72	0.101
Residual	30	1.38542	0.04618		
Total	47	2.58479			

Appendix Table 21.Analysis of variance of kernel yield (kg/ha) at Lemlem experimental site

Source of variation	d.f.	s.s.	m.s.	v.r.	F pr.
Rep stratum	2	424015.	212008.	7.28	
Rep.*Units* stratum					
mgt	15	935146.	62343.	2.14	0.037
Residual	30	873968.	29132.		
Total	47	2233129			

Appendix Table 22. Analysis of variance of 100 seed wt (gm) at Lemlem experimental site

Source of variation	d.f.	s.s.	m.s.	v.r.	F pr.
Rep stratum	2	0.650	0.325	0.09	
Rep.*Units* stratum					
mgt	15	339.793	22.653	6.17	<.001
Residual	30	110.196	3.673		
Total	47	450.640			

Appendix Table 23. Analysis of variance of shelling percentage (%) at Lemlem experimental site

Source of variation	d.f.	s.s.	m.s.	v.r.	F pr.
Rep stratum	2	0.016772	0.008386	2.39	
Rep.*Units* stratum					
mgt	15	0.043105	0.002874	0.82	0.651
Residual	30	0.105360	0.003512		
Total	47	0.165237			

Appendix Table 24. Analysis of variance of Harvest index (%) at Lemlem experimental site

Source of variation	d.f.	s.s.	m.s.	v.r.	F pr.
Rep stratum	2	0.042891	0.021446	5.39	
Rep.*Units* stratum					
mgt	15	0.073734	0.004916	1.24	0.300
Residual	30	0.119252	0.003975		
Total	47	0.235878			

Appendix Table 25. Analysis of variance of available soil moisture content (%) at 0- 40cm root depth at Lemlem experimental site

Source of variation	d.f.	s.s.	m.s.	v.r.	F pr.
Rep stratum	2	89.405	44.702	17.18	
Rep.*Units* stratum					
mgt	15	303.861	20.257	7.78	<.001
Residual	30	78.081	2.603		
Total	47	471.346			

Appendix Table26. Analysis of variance of available soil moisture content (%) at 41- 80cm

Source of variation	d.f.	s.s.	m.s.	v.r.	F pr.
Rep stratum	2	37.043	18.522	2.83	
Rep.*Units* stratum					
mgt	15	332.864	22.191	3.39	0.002
Residual	30	196.419	6.547		
Total	47	566.327			

Appendix Table 27. Analysis of variance of *Aspergillus niger* (transformed data) at Lemlem

Source of variation	d.f.	s.s.	m.s.	v.r.	F pr.
Rep stratum	2	10.1238	5.0619	12.19	
Rep.*Units* stratum					
mgt	15	6.8738	0.4583	1.10	0.394
Residual	30	12.4556	0.4152		
Total	47	29.4533			

Appendix Table 28. Incidence of *A.flavus* and *A.niger*

		Incidence (yes/no), 1= yes ,0= no			
		A.flavus		A.niger	
No.	trt	Hadinet	Lemelm	Hadinet	Lemelm
1	C	1	1	1	1
2	GP	1	1	1	1
3	SI	0	1	1	1
4	TD	1	1	1	1
5	F	1	1	1	1
6	GP +SI	1	1	1	1
7	GP +TD	1	1	1	1
8	GP + F	1	1	1	1
9	SI + TD	1	1	1	1
10	SI + F	1	1	1	1
11	TD + F	1	1	1	1
12	GP + SI +TD	1	1	1	1
13	GP +SI + F	1	1	1	1
14	GP +TD + F	1	1	1	1
15	SI +TD + F	1	1	1	1
16	SI +TD + F + GP	1	1	1	0
n/N*100		15/16*100	16/16*100	16/16*100	15/16*100
(%)		93%	100%	100%	93%

9 783659 379758